Statistical Methodology and Result Interpretation

Statistical Methodology and Result Interpretation

Editor: Isabel Young

STATES
ACADEMIC PRESS
www.statesacademicpress.com

Published by States Academic Press
109 South 5th Street,
Brooklyn, NY 11249, USA
www.statesacademicpress.com

Statistical Methodology and Result Interpretation
Edited by Isabel Young

International Standard Book Number: 978-1-63989-497-0 (Hardback)

Cataloging-in-Publication Data

Statistical methodology and result interpretation / edited by Isabel Young.
p. cm.
Includes bibliographical references and index.
ISBN 978-1-63989-497-0
1. Mathematical statistics. 2. Statistics--Methodology. I. Young, Isabel.
QA276 .S83 2022
519.5--dc23

Contents

Preface

I am honored to present to you this unique book which encompasses the most up-to-date data in the field. I was extremely pleased to get this opportunity of editing the work of experts from across the globe. I have also written papers in this field and researched the various aspects revolving around the progress of the discipline. I have tried to unify my knowledge along with that of stalwarts from every corner of the world, to produce a text which not only benefits the readers but also facilitates the growth of the field.

Statistics is a group of methods by which data is collected, organized and analyzed, and then its characteristics are interpreted and presented. Regression, mean, standard deviation, sample size determination and hypothesis testing are some of the methods used in statistical analysis. Statistics is applied in a large number of fields such as machine learning and data mining, statistical computing, business statistics, etc. From theories to research to practical applications, case studies related to all contemporary topics of relevance to this field have been included in this book. It discusses the fundamentals as well as modern approaches of statistical methodologies. As this field is emerging at a rapid pace, the contents of this book will help the readers understand the modern concepts and applications of the subject.

Finally, I would like to thank all the contributing authors for their valuable time and contributions. This book would not have been possible without their efforts. I would also like to thank my friends and family for their constant support.

Editor

1

Applications of the H-Principle of Mathematical Modelling

Agnar Höskuldsson

Abstract

Traditional statistical test procedures are briefly reviewed. It is pointed out that significance testing may not always be reliable. The author has formulated a modelling procedure, the H-principle, for how mathematical modelling should be carried out in the case of uncertain data. Here it is applied to linear regression. Using this procedure, the author has developed a common framework for carrying out linear regression. Six regression methods are analysed by this framework, two stepwise methods: principal component regression, ridge regression, PLS regression and an H-method. The same algorithm is used for all methods. It is shown how model validation and graphic analysis, which is popular in chemometrics, apply to all the methods. Validation of the methods is carried out by using numerical measures, cross-validation and test sets. Furthermore, the methods are tested by a blind test, where 40 samples have been excluded. It is shown how procedures in applied statistics and in chemometrics both apply to the present framework.

Keywords: linear regression, common framework for linear regression, stepwise regression, ridge regression, PLS regression, H-methods

1. Introduction

Regression analysis is the most studied subject within theoretical statistics. Numerous books have been published. Advanced program packages have been developed that make it easy for users to develop and study advanced models and methods.

Advanced program packages such as SAS and SPSS have been used by students since the 1970s and the program packages have become very advanced. The user fills out a 'menu' similar to those at restaurants. The program then carries out the analysis as requested providing the users' possibility of carrying out highly advanced analysis of the data. Users rely upon that they can make interpretation of the output in the way they learned at school. For example, if a variable/factor is significant, its presence in the model improves the predictions derived from the modelling results.

Today measurement instruments are becoming more and more advanced, e.g. optical instruments are becoming popular in industry. They may give thousands of values each time a sample is measured. In applied research, the tendency is to include as much information as possible in order to cover possible alternatives of the situation. When adding interactions or non-linear terms, we also see data having hundreds or thousands of variables.

There are three basic challenges in applied data analysis, which are as follows: (1) Data in applied sciences and industry are typically generated by instruments that produce many variables. In these cases, the data have **latent structure**, which means that the data values are geometrically located in a low-dimensional space. However, mathematical models and methods usually assume that the **X**-data have full rank. In these cases, the models are often incorrect and may give imprecise results.

(2) Typically, scientists develop solutions that are optimal or unbiased. This is a natural approach, because the derived solutions have important properties, when data satisfy the given model. Simulations based on the model confirm the optimality, but often in practice the data often do not satisfy the assumptions of the model. Forcing an optimal solution on the data may give results that have bad or no prediction ability. A better solution may then be obtained **by relaxing** on the optimality or unbiasedness of the solution.

(3) It is a tradition to base results of regression analysis on testing the significance of the variables in the model. However, the results may not always be reliable. The influence of a variable can be so small that it may be of no importance, see Section 3.3. At professional organizations, it is considered to be serious problem for the researchers in the field.

H-principle is a prescription of how a solution to a mathematical model should be generated, when data are uncertain. It is proposed that the determination of the solution should be carried out in steps and compute the improvement in the solution and the associated precision at each step. It is suggested that the solution at each step should be an optimal balance between the improvement and the associated precision. Determination of the solution continues as long as it is supported by the given (uncertain) data.

The author has developed a framework linear regression, which is inspired by the H-principle. Using this framework, the same algorithm is applied for carrying out different types of regression analysis. Most regression methods based on linear algebra can be carried out within this framework. Here, we carry out the analysis of six different methods. Numerical and graphic analyses of the results are the same for all the regression methods.

In Section 2, we specify the used notation and briefly describe the used data. In Section 3, we consider some basic issues in modelling data. The background for the H-principle is treated. In Section 4, we consider the latent variable regression model. In Section 5, we present the H-principle and show some examples of its usage. In Section 6, we present a common framework for linear regression. In Section 7, we discuss model validation. This is an important topic, but we only consider essential aspects that are used in the analysis of the six methods. In Section 8, we present the results for six different regression methods: (1) stepwise (forward) regression maximising covariance, (2) stepwise (forward) regression maximising R^2,

(3) principal component regression, (4) ridge regression,

(5) PLS regression and (6) an H-method.

Section 9 briefly discusses the results presented. In Section 10, we mention application of the H-principle to multi-block and path modelling, non-linear modelling and extension to multi-linear algebra. Section 11 presents conclusions.

2. Notation, data and scaling

Matrices and vectors are denoted by bold letters; matrices by upper case and vectors by lower case. In order to facilitate the reading of the equations, different types of indices are used for the steps and for the matrices/vectors. The letters *a* and *b* are related to the steps in the algorithm. The letters *i*, *j* and *k* are used for the indices within a matrix/vector. It is assumed that there is given instrumental data **X**, N × K matrix, and response data **Y**, N × M matrix. The regression data are denoted by (**X**, **Y**). In some cases, only one *y*-variable, **y**, is used. This is done to simplify the equations. It is assumed that data are centred, which means that average values are subtracted from each column of **X** and **Y**. This also makes it easier to read the equations. $\mathbf{x}_j$ is the jth column of **X** and $\mathbf{x}^i$ is the ith row of $\mathbf{X} = (x_{ij})$. A *latent variable* τ is a linear combination of the original measured variables,

$$\tau = a_1 x_1 + a_2 x_2 + \ldots + a_K x_K \tag{1}$$

The data used here are from an optical instrument. The instrument gives 1200 values at each measurement. However, technical knowledge of the instrument suggests that only 40 values should be used for determining the substance in question, the *y*-values. Two hundred samples are measured. Forty samples are put aside for a blind test. Thus, the calibration analysis is based on 160 samples. This gives **X** as 160 × 40 matrix and **y** a 160 vector. These data are challenging and represent a common situation, when working with optical instruments (FTIR, NIR, fluorescence etc.).

It is sometimes important to determine if data should be scaled or not. Scaling can be obtained by multiplying **X** and **Y** by diagonal matrices. If $\mathbf{C}_1$ and $\mathbf{C}_2$ are diagonal matrices, the scaling of **X** is done by the transformation, $\mathbf{X} \leftarrow (\mathbf{X}\mathbf{C}_1)$ and of **Y** by $\mathbf{Y} \leftarrow (\mathbf{Y}\mathbf{C}_2)$. The linear least squares solution for (**X**, **Y**) is given by $\boldsymbol{B} = (\boldsymbol{X}^T\boldsymbol{X})^{-1}\boldsymbol{X}^T\boldsymbol{Y}$. This solution can be obtained from the linear least squares solution for scaled data, $\mathbf{B}_1$, as follows,

$$\boldsymbol{B} = \boldsymbol{C}_1[\{(\boldsymbol{X}\boldsymbol{C}_1)^T(\boldsymbol{X}\boldsymbol{C}_1)\}^{-1}(\boldsymbol{X}\boldsymbol{C}_1)^T(\boldsymbol{Y}\boldsymbol{C}_2)]\boldsymbol{C}_2^{-1} = \boldsymbol{C}_1\boldsymbol{B}_1\boldsymbol{C}_2^{-1} \tag{2}$$

This shows that when computing the linear least squares solution, we can work with the scaled data. The original solution is obtained by scaling 'back' as shown in the equation. We use this property also, when we compute the approximate solution. The effect of scaling is better numerical precision. Scaling is necessary for the present data. If data are not scaled (e.g. to unit variance), numerical results beyond dimension of around 15 may not be reliable. The reason is that Eqs. (20) and (23) are sensitive to small numerical values. Scaling is much debated among researchers. When scaling is used, one must secure that all variables have values above the 'noise' level of the instrument. This is a difficult topic, which is not considered closer here. If original data follow a normal distribution, the scaled ones will not. However, for large sample number like given here, the differences will be negligible.

3. Linear regression

3.1. Traditional regression model

It is assumed that there are given instrumental data **X** and response data **y**. A linear regression model is given by

$$y = \beta_1 x_1 + \beta_2 x_2 + \ldots + \beta_K x_K + \varepsilon \tag{3}$$

The x-variables are called independent variables and the y-variable is the dependent one. When the parameters $\boldsymbol{\beta}$ have been estimated as **b**, the estimated model is now

$$y = b_1 x_1 + b_2 x_2 + \ldots + b_K x_K \tag{4}$$

When there is given a new sample $\mathbf{x}_0 = (x_{10}, x_{20}, \ldots, x_{K0})$, it gives the estimated or predicted value $y_0 = b_1\, x_{10} + \ldots + b_K\, x_{K0}$. There can be many estimates for the regression coefficients. Therefore, Greek letters are used for the theoretical parameters and Roman letters for the estimated values. It is common to use the linear least squares method for estimating the parameters. It is based on minimizing the residuals, $(\mathbf{y}-\mathbf{X}\boldsymbol{\beta})^T(\mathbf{y}-\mathbf{X}\boldsymbol{\beta}) \rightarrow$ minimum, with respect to $\boldsymbol{\beta}$. The solution is given by

$$\boldsymbol{b} = (\boldsymbol{X}^T\boldsymbol{X})^{-1}\boldsymbol{X}^T\boldsymbol{y}. \tag{5}$$

It is common to assume that the residuals in Eq. (3) are normally distributed. This is often written as $\mathbf{y}\sim N(\mathbf{X}\boldsymbol{\beta}, \sigma^2)$. It means that the expected value of **y** is $E(\mathbf{y}) = \mathbf{X}\boldsymbol{\beta}$ and the variance is $Var(\mathbf{y}) = \sigma^2\mathbf{I}$, where **I** is the identity matrix. The linear least squares procedure coincides with the maximum likelihood method in case the data follow a normal distribution. Assuming the normal distribution, the parameter estimate **b** has the variance given by

$$Var(b) = \sigma^2(X^TX)^{-1}. \tag{6}$$

Here σ^2 is estimated by

$$\sigma^2 \cong \Sigma_1^N e_i^2/(N-K) \tag{7}$$

where the residuals, e_is, are computed from $\mathbf{e} = \mathbf{y} - \mathbf{Xb}$.

3.2. Assumptions and properties

It is assumed that the matrix $\mathbf{X}^T\mathbf{X}$ has an inverse. Furthermore, it is assumed that the samples are independent of each other. In case the model (Eq. (3)) is correct and data follow the normal distribution, the estimate (Eq. (5)) has some important properties compared with other possible estimates:

a. It is unbiased, $E(\mathbf{b}) = \boldsymbol{\beta}$.

b. The estimate (Eq. (5)) has among all linear unbiased estimates of $\boldsymbol{\beta}$ the smallest variance. It means that if $\mathbf{b}_1$ is another unbiased estimate of $\boldsymbol{\beta}$, then $Var(\mathbf{b}_1) - Var(\mathbf{b})$ is a semi-positive definite matrix.

There is a general agreement that the variance matrix $Var(\mathbf{b})$ should be as small as possible. The matrix $(\mathbf{X}^T\mathbf{X})^{-1}$, the *precision matrix*, shows how precise the estimates $\mathbf{b}$ are. The properties (a) and (b) state that assuming the linear model (Eq. (3)), the solution (Eq. (5)) is, in fact from a theoretical point of view, the best possible one. It often occurs that the model (Eq. (3)) contains many variables. Interpretation of the estimation results (Eq. (4)) is an important part of the statistical analysis. Therefore, it is common to evaluate the parameters with the aid of a significance test. If a variable is not significant, it may be excluded. A parameter associated with a variable is commonly evaluated by a *t*-test. A *t*-test of a parameter b_i is given by

$$t = \frac{b_i}{\sqrt{Var(b_i)}} \cong \frac{b_i}{\sqrt{(s^2 \times s^{ii})}} \tag{8}$$

Here s^2 is given by Eq. (7) and s^{ii} is the *i*th diagonal element of $(\mathbf{X}^T\mathbf{X})^{-1}$.

This is the motivation that the program packages such as SAS and SPSS in statistics compute the linear least squares solution in the case of linear regression analysis as the initial solution. Significance of the parameters in Eq. (4) is shown using Eq. (8).

The problem in using the least squares solution appears when the precision matrix becomes close to singular. If the computer program (SAS or SPSS) detects that the computation of $\mathbf{b}$ may be imprecise due to close to singularity of the precision matrix, the user is informed that the estimates $\mathbf{b}$ may be imprecise. However, the information is related to the judgement of the situation and the numerical precision of the computer. There are practical issues long before the question of the numerical uncertainties of the estimates, $\mathbf{b}$'s.

Consider two examples. Suppose that $\mathbf{X}$ is $N \times 2$ and that $\mathbf{x}_1 = c\,\mathbf{x}_2 + \boldsymbol{\delta}$, for some constant c, where $|\boldsymbol{\delta}| < 10^{-5}$. We may be able to compute $(\mathbf{X}^T\mathbf{X})^{-1}$. However, inference on the model (Eq. (4)) may be uncertain or incorrect. At the other example, suppose that $\mathbf{X}$ is $N \times 40$, but that practical rank is 15. Assume that $\mathbf{X}^T\mathbf{X} = \mathbf{X}_1^T\mathbf{X}_1 + \mathbf{X}_2^T\mathbf{X}_2$, where $\mathbf{X}_2 = (\mathbf{x}_{2,1}\ \mathbf{x}_{2,2} \ldots \mathbf{x}_{2,40})$. If all $\mathbf{x}_{2,i}$ $i = 1,\ldots,40$ are small, say $|\mathbf{x}_{2,i}| < 10^{-5}$, inference from Eq. (4) may be uncertain or incorrect. The first example may not be realistic. However, the second one is realistic. It is common that data in applied sciences and industry are of reduced practical rank. In these cases, the model (Eq. (3)) is incorrect and leads to uncertain or incorrect results.

3.3. Stepwise linear regression

We shall consider closer the procedure of stepwise linear regression. We do that by using the Cholesky factorization $\mathbf{X}^T\mathbf{X} = \mathbf{F}\mathbf{F}^T$, where $\mathbf{F}$ is lower triangular. Then we can write the columns of the data matrix $\mathbf{X}$ as

$$\begin{aligned} x_1 &= F_{11}t_1 \\ x_2 &= F_{21}t_1 + F_{22}t_2 \\ &\dots \\ x_K &= F_{K1}t_1 + F_{K2}t_2 + \dots + F_{KK}t_K \end{aligned} \tag{9}$$

The vectors ($\mathbf{t}_i$) are mutually orthogonal and of length 1, $\mathbf{t}_i^T\mathbf{t}_j = \delta_{ij}$. The significance of $\mathbf{x}_K$, when $\mathbf{x}_1, \mathbf{x}_2, \dots, \mathbf{x}_{K-1}$ are given, is computed from Eq. (8) with

$$b_K = \frac{y^T(F_{KK}t_K)}{F_{KK}^2} = \frac{(y^Tt_K)}{F_{KK}} \ and\ s^{KK} = 1/F_{KK}^2 \tag{10}$$

This gives

$$t = \frac{(y^Tt_K)}{s} \tag{11}$$

This shows that the *t*-test for the significance of $\mathbf{x}_K$ is independent of the size F_{KK}. When the selection of a new variable is carried out among many variables (e.g. 500), there is a considerable risk that F_{KK} is so small that the marginal effect of $\mathbf{x}_K$ is of no importance although it is being declared as statistically significant. The issue is that the user of program packages is not informed of, if a significant variable improves the prediction of the response variable or not.

3.4. Variance of regression coefficients

It is instructive to study closer the variance of the regression coefficient in the linear least squares model, Eq. (6). It can be written as

$$Var(\boldsymbol{b}) \cong [y^Ty - y^TX(X^TX)^{-1}X^Ty] \times [(X^TX)^{-1}]/(N-K) \tag{12}$$

An important objective of the modelling task is to keep the value of Eq. (12) as small as possible. Equation (12) is a product of two parts. For these two parts it can be shown that, when a new variable is added to the model

i. the residual error, $[y^Ty - y^TX(X^TX)^{-1}X^Ty]$, always decreases

ii. the precision matrix, $(X^TX)^{-1}$, always increases

(In theory these measures can be unchanged, but in practice these changes always occur). Note that the two terms, (i) and (ii), are equally important and appear in a symmetric way in Eq. (12). If the data are normally distributed, it can be shown that

a. the residual error, $[y^Ty - y^TX(X^TX)^{-1}X^Ty]$,

b. the precision matrix, $(X^TX)^{-1}$,

are stochastically independent. This means that the knowledge of one of them does not give any information on the other one. If, for instance, a certain fit has been obtained, then there is no information on (b). In order to know the quality of predictions, we must be compute (b) and use Eq. (12). Knowing (a), the results concerning the precision matrix can be good or bad. Program packages only use the *t*-test or some equivalent measure to evaluate to test significance of the variables/components. There is no information on the precision matrix. We must compute the precision matrix together with the significance testing in order to find out how well the model is performing. **However, a modelling procedure *must* include both (a) and (b) in order to secure small values of Eq. (12).** The procedure must handle both the decrease of (i) and the increase of (ii) during the model estimation.

4. Latent variable model

The measurement data in applied sciences and industry often represent a 'system', where variables are dependent on each other from chemical equilibrium, action-reaction in physics and physical and technical balance. Geometrically, the samples are located in a low-dimensional space. In these cases, the model (Eq. (3)) is **incorrect** (apart from simple, designed laboratory experiments) and the nice theory above may not applicable. Why is this the case? When a new sample is available, we cannot automatically insert it into Eq. (4) and compute the y-value. We must secure that the new sample is located geometrically along the samples used for analysis, the rows of **X**. A correct model for this kind of data is

$$y = \alpha_1\tau_1 + \alpha_2\tau_2 + \ldots + \alpha_A\tau_A + \varepsilon. \tag{13}$$

Here (τ_a) are *latent variables* and (α_a) are the regression coefficients on the latent variables. The regression task is both to determine the latent variables, τ's, and to compute the associated regression coefficients. The resulting Eq. (13) is then converted into Eq. (4) using Eq. (26). The model (Eq. (13)) is called **latent structure** linear regression. It is sometimes difficult to make 'good' interpretation of the variables, when using latent variables. However, by using appropriate modelling procedure, it is possible to obtain a fairly good interpretation of the variables (see Section 5.4).

5. The H-principle of mathematical modelling

5.1. Background

In the 1920s, there were large discussions on the measurement aspects of quantum mechanics. W. Heisenberg pointed out that there are certain magnitudes that cannot be determined exactly at the same time. His famous uncertainty inequality states that there is a lower limit to how well conjugate magnitudes can be determined at the same time. The position and momentum of an elementary particle is an example. For that example, the inequality is

$$\Delta(\text{position}) \times \Delta(\text{momentum}) \geq \text{constant} \tag{14}$$

The lower limit of the inequality is related to the Planck's constant of light. These considerations are based on physical theory. In practice, it means that there are some restrictions on the outcome of an experiment. The results may depend on the instrument and the phenomenon being studied. The restrictions are detected by the application of the measurement instrument. The uncertainty inequality and the associated theory are a kind of guidance, when carrying out experiments.

5.1.1. The H-principle

When modelling data, there is an analogous situation. Instead of a measurement instrument we have a mathematical method. The conjugate magnitudes are ΔFit and ΔPrecision and they cannot be controlled at the same time. It is recommended to carry out the modelling in steps and at each step evaluate the situation as prescribed by the uncertainty inequality. At each step, it is assumed that there is given a weight vector. The recommendations of the H-principle of mathematical modelling are the following (expressed in the case of regression analysis):

1. Carry out determining the solution in steps. You specify how you want to look at the data at this step by formulating how the weights are computed.
2. At each step compute
   a. expression for improvement in fit, Δ(Fit)
   b. the associated prediction, Δ(Precision)
3. Compute the solution that minimises the product Δ(Fit) × Δ(Precision)
4. In case the computed solution improves the prediction abilities of the model, the solution is accepted. If the solution does not provide this improvement, the modelling stops.
5. The data are adjusted for what has been selected; restart at 1.

Note that the H-principle is a recommendation of how to proceed in determining the solution of a mathematical problem, when data are uncertain. It is not like maximum likelihood, where a certain function is to be maximized. However, it is necessary to compute the improvement in the solution at each step and the associated precision. The optimal balance is described in Section 3 and the situation is evaluated to find out if data are following along.

5.1.2. Application to linear regression

Let us consider closer how this applies to linear regression. The task is to determine a weight vector $\mathbf{w}$ according to this principle. For the score vector $\mathbf{t} = \mathbf{X}\mathbf{w}$ we have

a. Improvement in fit: $|\boldsymbol{Y}^T\boldsymbol{t}|^2/(\boldsymbol{t}^T\boldsymbol{t})$

b. Associated variance: $\Sigma/(\boldsymbol{t}^T\boldsymbol{t})$

If $\mathbf{S} = \mathbf{X}^T\mathbf{X}$ the adjustment (Eq. (20)) gives orthogonal score vectors, $\mathbf{t}$'s. Therefore, the variance, (*b*), is used for the precision. Treating $\boldsymbol{\Sigma}$ as a constant the task is to maximize

$$\left[\frac{|Y^T t|^2}{t^T t}\right] \times \frac{1}{\left[\frac{1}{t^T t}\right]} = w^T X^T Y Y^T X w \tag{15}$$

This is a maximization task because improvement in fit is negative. The maximization is carried out under the restriction that **w** is of length 1, $|\mathbf{w}|=1$. By using the Lagrange multiplier method, it can be shown that the task is an eigenvalue task,

$$X^T Y Y^T X w = \lambda w \tag{16}$$

In case there is only one y-variable, the eigenvalue task has a direct solution

$$w = \frac{X^T y}{|X^T y|} = \frac{C}{|C|} \tag{17}$$

These are the solutions used in PLS regression. Thus, we can state that the PLS regression is consistent with the H-principle. H-methods take as a starting point the PLS solution and determine how the prediction aspect or the solution can be improved.

5.2. Interpretation of the modelling results

The fit obtained by the score vector **t** is

$$\frac{(y^T t)^2}{(t^T t)} = \frac{(C^T C)^2}{C^T S C} = \left[\frac{c_1^2}{f} + \frac{c_2^2}{f} + \ldots + \frac{c_K^2}{f}\right]^2 \tag{18}$$

where $f = \sqrt{C^T SC}$

The regression coefficient can be written similarly as

$$\frac{(y^T t)}{(t^T t)} = \frac{(C^T C)^{1.5}}{C^T S C} = \left[\frac{c_1^2}{g} + \frac{c_2^2}{g} + \ldots + \frac{c_K^2}{g}\right]^{1.5} \tag{19}$$

where $= (C^T SC)^{\frac{1}{1.5}}$.

At each step, we can see how much each variable contributes to the fit and the regression coefficient..

6. A common framework for linear regression

6.1. Background: views on the regression analysis task

There are many ways to carry out a regression analysis. Here we present a general framework for carrying out linear regression that includes most methods based on linear algebra. The basic idea is to separate the computations into two parts: the first part is concerned on how it is preferable to look at data. In practice, there can be different emphasis on the regression analysis. Sometimes it is important to determine important variables. In other cases, it may be the predictions derived that is important. The other part is the numerical algorithm to compute the solution vector and associated measures. At the first part, it is assumed that a weight matrix $\mathbf{W} = (\mathbf{w}_1, \ldots, \mathbf{w}_K)$ is given. Each weight vector should be of length one, although this is not necessary. The weight vectors ($\mathbf{w}_a$) specify how one wants to look at the data. They may be determined by some optimization or significance criteria. They may also be determined by a standard regression method such as PLS regression. In Section 6.4, we show the choices in the case of the six regression analysis. The role of the weight vectors is only to compute the loading vectors. The other part of the computations is a numerical algorithm, which is the same for all choices of **W**. There is no restriction on the weight vectors except that they may not give zero-loading vector, $\mathbf{p}_a \neq \mathbf{0}$.

6.2. Decomposition of data

The starting point is a variance matrix **S** and a covariance matrix **C**. **S** can be any positive semi-definite matrix, but **C** is assumed to be the covariance, $\mathbf{C} = \mathbf{X}^T\mathbf{Y}$. The algorithm is formulated for multiple y's.

Initially $\mathbf{S}_0 = \mathbf{S}$ and $\mathbf{C}_0 = \mathbf{C}$ and $\mathbf{B} = \mathbf{0}$. For $a = 1, \ldots, K$:

$$\text{Loading vector}: \boldsymbol{p}_a = \boldsymbol{S}_{a-1}\boldsymbol{w}_a \tag{20}$$

$$\text{Scaling constant}: d_a = \frac{1}{\boldsymbol{w}_a^T\boldsymbol{p}_a} \tag{21}$$

$$\text{Loading vector}: \boldsymbol{q}_a = \boldsymbol{C}_{a-1}^T\boldsymbol{w}_a \tag{22}$$

$$\text{Loading weight vector}: \boldsymbol{v}_a \tag{23}$$

The loading weight vectors are computed by, see reference [1],

$$\boldsymbol{v}_1 = \boldsymbol{w}_1, \boldsymbol{v}_a = \boldsymbol{w}_a - d_1(\boldsymbol{p}_1^T\boldsymbol{w}_a)\boldsymbol{v}_1 - \ldots - d_{a-1}(\boldsymbol{p}_{a-1}^T\boldsymbol{w}_a)\boldsymbol{v}_{a-1}, a = 2, 3, \ldots, K \tag{24}$$

S is adjusted by the loading vector $\mathbf{p}_a$ and similarly for **C**,

$$\boldsymbol{S}_a = \boldsymbol{S}_{a-1} - d_a\boldsymbol{p}_a\boldsymbol{p}_a^T \tag{25}$$

$$\boldsymbol{C}_a = \boldsymbol{C}_{a-1} - d_a\boldsymbol{p}_a\boldsymbol{q}_a^T \tag{26}$$

The adjustment of **S** is of rank one reduction. **S** also reduces in size by

$$tr(d_a\boldsymbol{p}_a\boldsymbol{p}_a^T) = d_a\boldsymbol{p}_a^T\boldsymbol{p}_a = \frac{\boldsymbol{w}_a^T\boldsymbol{S}^2\boldsymbol{w}_a}{\boldsymbol{w}_a^T\boldsymbol{S}\boldsymbol{w}_a} > 0. \tag{27}$$

The loading weight matrix **V** satisfies $\mathbf{V}^T\mathbf{P} = \mathbf{D}^{-1}$, see reference [1].

6.3. Expansion of matrices

At $a = K$ we get $\mathbf{S}_K = \mathbf{0}$. Expanding **S** and **C** we get

$$\boldsymbol{S} = d_1\boldsymbol{p}_1\boldsymbol{p}_1^T + \ldots + d_A\boldsymbol{p}_A\boldsymbol{p}_A^T + \ldots + d_K\boldsymbol{p}_K\boldsymbol{p}_K^T = PDP^T \tag{28}$$

$$C = \boldsymbol{X}^T\boldsymbol{Y} = d_1\boldsymbol{p}_1\boldsymbol{q}_1^T + \ldots + d_A\boldsymbol{p}_A\boldsymbol{q}_A^T + \ldots + d_K\boldsymbol{p}_K\boldsymbol{q}_K^T = \boldsymbol{PDQ}^T \tag{29}$$

Inserting appropriate matrices we get

$$\boldsymbol{S}^{-1} = d_1\boldsymbol{v}_1\boldsymbol{v}_1^T + \ldots + d_A\boldsymbol{v}_A\boldsymbol{v}_A^T + \ldots + d_K\boldsymbol{v}_K\boldsymbol{v}_K^T = \boldsymbol{VDV}^T \tag{30}$$

$$\boldsymbol{B} = \boldsymbol{S}^{-1}X^TY = d_1\boldsymbol{v}_1\boldsymbol{q}_1^T + \ldots + d_A\boldsymbol{v}_A\boldsymbol{q}_A^T + \ldots + d_K\boldsymbol{v}_K\boldsymbol{q}_K^T = \boldsymbol{VDQ}^T \tag{31}$$

$$\widehat{Y}^T\widehat{Y} = d_1\boldsymbol{q}_1\boldsymbol{q}_1^T + \ldots + d_A\boldsymbol{q}_A\boldsymbol{q}_A^T + \ldots + d_K\boldsymbol{q}_K\boldsymbol{q}_K^T = \boldsymbol{QDQ}^T \tag{32}$$

If $\mathbf{S} = \mathbf{X}^T\mathbf{X}$, we can expand **X** and **Y** in a similar way. Compute a score matrix **T** by $\mathbf{T} = \mathbf{XV}$. This gives

$$\boldsymbol{X} = d_1\boldsymbol{t}_1\boldsymbol{p}_1^T + \ldots + d_A\boldsymbol{t}_A\boldsymbol{p}_A^T + \ldots + d_K\boldsymbol{t}_K\boldsymbol{p}_K^T = \boldsymbol{TDP}^T \tag{33}$$

$$\widehat{Y} = d_1\boldsymbol{t}_1\boldsymbol{q}_1^T + \ldots + d_A\boldsymbol{t}_A\boldsymbol{q}_A^T + \ldots + d_K\boldsymbol{t}_K q_K^T = \boldsymbol{TDQ}^T \tag{34}$$

The score vectors are mutually orthogonal. This follows from $\mathbf{T}^T\mathbf{T} = \mathbf{D}\text{–}^1$. Generally, only A terms of the expansions are used, because it is verified that the modelling task cannot be improved beyond A terms. For the proof of geometric properties of the vectors in these expansions, see reference [1].

It can be recommended to use a test set, ($\mathbf{X}_t$, $\mathbf{Y}_t$), when carrying out a regression analysis. Centring (and scaling if used) is done on the test set by using the corresponding values from (**X**, **Y**). The estimated y-values are computed as $\hat{\mathbf{Y}}_t = \mathbf{X}_t\,\mathbf{B}$ and the score vectors for the test set as $\mathbf{T}_t = \mathbf{X}_t\,\mathbf{V}$. It is often useful to study the plots, showing columns of **Y** against columns of **T** (in stepwise regression, the plots are called added variable plots). Similarly, we can plot the score vectors of the test set, the columns of $\mathbf{T}_t$, against columns of $\mathbf{Y}_t$.

6.4. Choices of weight vectors

The weight vectors for the six regression methods are as follows:

(i) Stepwise regression, maximize covariance:

$$w_a(ii) = 1 \text{ for } C_{ii} = \max_{i=1}^{K}|C_i| \text{ and } = 0 \text{ otherwise} \tag{35}$$

(ii) Stepwise regression, maximize R^2:

$$w_a(ii) = 1 \text{ for} \frac{(\boldsymbol{y}^T\boldsymbol{x}_{ii})^2}{(\boldsymbol{x}_{ii}^T\boldsymbol{x}_{ii})} = \max_{i=1}^{K}\frac{(\boldsymbol{y}^T\boldsymbol{x}_i)^2}{(\boldsymbol{x}_i^T\boldsymbol{x}_i)}, \text{ and } = 0 \text{ otherwise} \tag{36}$$

(iii) and (iv) Eigenvector of S:

$$\text{For } \boldsymbol{S} = \boldsymbol{UEU}^T, \ \boldsymbol{w}_a = \boldsymbol{u}_a \tag{37}$$

(v) PLS regression:

$$\boldsymbol{w}_a = \boldsymbol{C}_a / |\boldsymbol{C}_a| \tag{38}$$

where $\mathbf{C}_a$ is the reduced covariance.

(vi) H-method:

The weight vector of PLS regression is sorted with largest first, $\left(w_1^{(s)}, w_2^{(s)}, \ldots, w_K^{(s)}\right)$. Columns of $\mathbf{X}$ are rearranged to match this sorting, $\mathbf{X}^{(s)}$.

$$\boldsymbol{w}_{a,m} = \left(w_1^{(s)}, w_2^{(s)}, \ldots, w_m^{(s)}, 0, \ldots, 0\right) \tag{39}$$

The index m is chosen, which gives the best explained variation,

$$\boldsymbol{t}_i = \boldsymbol{X}^{(s)} \boldsymbol{w}_{a,i}, \ \frac{(\boldsymbol{y}^T \boldsymbol{t}_m)^2}{(\boldsymbol{t}_m^T \boldsymbol{t}_m)} = \max_{i=1}^{K} \frac{(\boldsymbol{y}^T \boldsymbol{t}_i)^2}{(\boldsymbol{t}_i^T \boldsymbol{t}_i)} \tag{40}$$

For further details of this method, see reference [2]. It has been applied to different bio-assay studies, where there can be many variables (3000 or more). Comparisons with several other methods have been shown that this method is preferable to work within the bio-assay studies, see references [3, 4]. Several other methods to improve the PLS solution have been developed.

7. Model validation

7.1. Numerical measures

The Mallow's Cp value and Akaike's information measure are commonly presented as results in a regression analysis. Mallow's Cp value is given by

$$C_p = \frac{|y - \hat{y}_A|^2}{\sigma^2} + 2A - N \tag{41}$$

As σ^2 we use the residual variance, s_A^2, at the maximal number of steps in the algorithm. C_p is an estimation of '(total means squared error)/σ^2'. The interpretation of C_p is

i. it should be as small as possible.

ii. its value should be as close to A as possible

iii. deviations of C_p from A suggests bias

Akaike's information measure is given by

$$AIC_A = N(\log(s_A^2) + 1) + s(A + 1) \tag{42}$$

It is an information measure that states the discrepancy between the correct model and the one obtained at step A. The number of components, A, is chosen that gives the smallest value of Eq. (32).

Both C_p and AIC have the property that they are not dependent on the given linear model, [5, 6]. Therefore, they can be used for all six methods considered here.

A *t*-value for the significance of a regression coefficient is given by

$$t\text{–value} = \frac{(\boldsymbol{y}^T \boldsymbol{t}_A)}{s_A} \tag{43}$$

Here $\mathbf{t}_A$ is the *A*th score vector of unit length. The significance, *p*-value, of the *t*-value can be computed using the *t*-distribution. Although the assumptions of a *t*-test are not valid for any of the six methods, it is useful to look at the significance. One can show that theoretically this value should be larger than 2 in order to be significant.

When comparing methods, it is useful to look at the estimate for the variance, *Var*(**b**), of the regression coefficients. We compute

$$(\text{trace}(Var(\boldsymbol{b}_A)))^{½} = (s_A^2 \textstyle\sum_{a=1}^{A} d_a (\boldsymbol{v}_a^T \boldsymbol{v}_a))^{½} \tag{44}$$

We cannot use this measure to determine the dimension. However, it is useful in comparing different methods.

7.2. The covariance

Eq. (15) is equivalent to the singular value decomposition of the covariance **C**. Therefore, it is suggested that the modelling of data should continue as long as the covariance is not zero. The dimension of a model can be determined by finding, when $\mathbf{C} = \mathbf{X}^T\mathbf{y} \cong \mathbf{0}$ for reduced matrices. One procedure is to study the individual terms, $(\mathbf{x}_i^T\mathbf{y})$. Assume that data can be described by a multivariate normal distribution with a covariance matrix $\mathbf{\Sigma}_{xy}$. Then, it is shown in reference [7] that the sample covariances $(\mathbf{x}_i^T\mathbf{y})/(N-3)$ are approximately normally distributed. If $\sigma_{xi,y} = 0$, it is shown in reference [7] that approximate 95% limits for the residual covariance, $(\mathbf{x}_i^T\mathbf{y})/(N-3)$, are given by

$$\pm 1.96\sqrt{N}\sigma_{xi}\sigma_y/(N-3) \cong \pm 1.96\sqrt{N}s_{xi}s_y/(N-3) \tag{45}$$

Thus, when modelling stops, it is required that all residual covariances should be within these limits. If $\sigma_{xi,y} = 0$, the distribution of the residual covariance approaches quickly the normal distribution by the central limit theorem. Therefore, it is a reliable measure to judge, if the covariances have become zero or close to zero.

Another approach is to study the total value, $\mathbf{y}^T\mathbf{X}\mathbf{X}^T\mathbf{y} = \Sigma_i(\mathbf{x}_i^T\mathbf{y})^2$. If the covariance matrix $\mathbf{\Sigma}_{xy}$ is zero, $\mathbf{\Sigma}_{xy} = \mathbf{0}$, the mean, $\mu = E\{(\mathbf{y}^T\mathbf{X}\mathbf{X}^T\mathbf{y})\}$, and variance, $\sigma^2 = Var\{(\mathbf{y}^T\mathbf{X}\mathbf{X}^T\mathbf{y})\}$, can be computed [8]. If the covariance is not zero, $\mathbf{\Sigma}_{xy} \neq \mathbf{0}$, it can be shown that $E\{(\mathbf{y}^T\mathbf{X}\mathbf{X}^T\mathbf{y})\} > \mu$. The upper 95% limit of a normal distribution $N(\mu,\sigma^2)$ is $\mu + 1.65\sigma$. This is used for mean and variance. In the analysis, it is checked if $\mathbf{y}^T\mathbf{X}\mathbf{X}^T\mathbf{y}$ is below the upper 95% limit (a one-sided test)

$$y^T XX^T y < \text{trace}(X^T X)(y^T y)/N + 1.65\sqrt{2\text{trace}(X^T XX^T X)(y^T y)^2/N^2} \tag{46}$$

When this inequality is satisfied, there is an indication that modelling should stop.

In the analysis in Section 8, we use the p-value,

$$p\text{-value} = P\{y^T XX^T y > \mu | N(\mu, \sigma^2)\} \tag{47}$$

This analysis has been found useful for different types of regression analysis.

7.3. Cross-validation and test sets

In stepwise regression, a search is carried out to find the next best variable. In PLS regression, a weight vector $\mathbf{w}$ is determined that gives the maximal size of the y-loading vector $\mathbf{q}$. When search or optimization procedures are carried out there is a considerable risk of overfitting. Evaluation of the results by standard statistical significance tests may show high significance, while other measures may show that the results are not significant. The uncertainties of predictions of new samples depend on the location of the samples in question. The further away from the centre (average values), the larger the uncertainties are. On the other hand, it is important to have large values, both large x-samples and y-values, in order to obtain stable estimates. In chemometrics, these special features of prediction are well known. A 10-fold cross-validation is often used. The samples are divided randomly into 10 groups. Nine groups are used for calibration (modelling) and the results applied to the 10th group. This is repeated for all groups so that all y-values are computed by a model, $\mathbf{y}_c$, that is using 90% of the samples. Experience has shown that this procedure may not always function well. As an example, one can mention the case, where the y-values have very skew distribution. There can be big difference in the results from cross-validation depending on how well large y-values are represented in the groups. It may be necessary that each group has a similar 'profile' as the total set of samples. For the present data, the y-values show very skew distribution (log(log(y-values)) show a normal distribution). It may be necessary that each group is representative for the whole data. This can be achieved by *ordered cross-validation*. Here, the samples are ordered in some way, which reflects the variation or sizes in data. In a 10-fold ordered cross-validation, the first group of samples is number 1, 11, 21, etc. of the ordered samples. The second group of samples is number 2, 12, 22, etc. In this way, each of the 10 groups of samples is representative with respect to the chosen ordering. The cross-validation is now carried out in the usual way.

As a result of cross-validation, we compute

$$D_a = 1 - |y - y_c|^2 / |y|^2 \; for \; a = 1, 2, \ldots, A \tag{48}$$

D_a is computed for each dimension. In the analysis below sorted y-values are used in the ordered cross-validation. A test set for the analysis is also selected in a similar way. Samples are ordered according to the first PLS score vector. 15% of the samples or $160 \times 0.15 = 24$ samples are used in the test set ($\mathbf{X}_t$, $\mathbf{y}_t$). Thus, the analysis in Section 8 is based on 136 samples. Equation (38) is also used, when applying results to a test set ($\mathbf{y} = \mathbf{y}_t$ and $\mathbf{y}_c = \mathbf{X}_t \times \mathbf{b}$).

8. Results for six different methods

8.1. Preliminary remarks

We have chosen here data that are challenging to work with. They are representative for many situations that industry and research projects find themselves in, when working with optical instruments. It is expected that large dimension is needed to model the data. The aim of the present work is to show how commonly used measures in applied statistics and popular chemometric procedures fit within the present framework. Therefore, we have chosen data, where there is not big difference between the six methods chosen here. We show that the different measures/procedures do not agree on what dimension should be used. Thus, the data analyst is choosing one set of measures that will typically be based on experience. It is a part of industry standard, see reference [9], to develop a calibration method. When the development work is finished, a blind test is carried out for 40 new samples. The data analyst should not know the new 40 reference values, y-values. The estimate for the new 40 samples should be evaluated by another person. This procedure is used here: the last 40 samples are put aside and the calibration analysis will be based on the other samples. When analysis has been completed for the six methods, we apply the results to the 40 samples. Optical instruments may give thousands of values for each measurement that is carried out (the one used here gives 1200 values). Experts often point out which part of the data should be used. Here it has been proposed to work with 40 variables, which all show significant correlation with the y-values. An upper limit for the dimension is chosen here to be 20. There is not a numerical problem in computing the precision matrix. The ratio between the largest and the smallest eigenvalue of $\mathbf{S}$ is $\lambda_1/\lambda_{40} = 9.0 \times 10^6$. However, experience suggests that the dimension should be less than 20.

8.2. Modelling results for six methods

In order to make readings of the following tables easier, we shall use the following headings for the tables.

i. C_p, Eq. (31), where σ^2 is the residual variance at dimension 20. For dimension 20, we get $C_p = 19$

ii. AIC, Eq. (32), where s_A^2 is the residual variance at dimension A

iii. the t-value Eq. (33), where s_A is the residual standard deviation at dimension A

iv. two-sided p-value for the t-test

v. the size of covariance $|\mathbf{X}^T\mathbf{y}|$ at dimension A

vi. one-sided p-value for the significance of (v), see Eq. (37)

vii. the estimate (Eq. (34)) for $\sqrt{\text{trace}(Var(\mathbf{b}_A))}$

viii. ordered cross-validation, samples ordered by y-values, Eq. (38)

ix. test set results, samples ordered by values of the first PLS score vector (Eq. (38)).

The improvement in fit, size of score vectors, etc. are not shown in the tables below. Only dimensions from 10 to 20 are shown. Dimensions from 1 to 9 are all significant for all methods.

8.2.1. Forward stepwise regression, maximal covariance

Here a variable is selected at each step that has maximal covariance $|(\mathbf{x}_i^T\mathbf{y})|$ for the reduced data (**Table 1**).

The 15th variable selection has the smallest AIC value and highly significant t-value. Therefore, it seems reasonable to choose 15 variables. Cross-validation and test set have reached the high level at dimension 10 or 11. This indicates some overfitting by choosing 15 variables.

8.2.2. Forward stepwise regression, maximal R^2-value

Here a variable is selected at each step that has maximal size of $(\mathbf{x}_i^T\mathbf{y})^2/(\mathbf{x}_i^T\mathbf{x}_i)$ for the reduced data (**Table 2**).

It seems appropriate to choose 16 variables. Seventeenth is at the boundary of being also significant. C_p for 16 variables indicates some bias present. Cross-validation and test set have reached high values at step 10, which indicates some overfitting.

8.2.3. Principal component regression

Here score vectors are selected that correspond to the associated eigenvalues (**Table 3**).

Score vector no 14 is not significant, and perhaps should be excluded. It seems appropriate to choose dimension 17 here. It is common to remove score vectors that are not significant.

No.	(i)	(ii)	(iii)	(iv)	(v)	(vi)	(vii)	(viii)	(ix)
10	90.5	−1571.9	12.65	0.0000	0.0518	0.000	0.23	0.9577	0.9761
11	88.0	−1572.2	−1.78	0.0765	0.0186	0.033	0.24	0.9609	0.9747
12	81.4	−1575.7	2.52	0.0127	0.0121	0.000	0.27	0.9640	0.9744
13	62.1	−1590.0	4.10	0.0001	0.0111	0.000	0.31	0.9631	0.9760
14	29.9	−1618.3	−5.61	0.0000	0.0070	0.000	0.43	0.9626	0.9804
15	**18.3**	**−1629.1**	**3.64**	**0.0003**	**0.0053**	**0.002**	**0.46**	**0.9658**	**0.9786**
16	19.1	−1627.3	−1.09	0.2772	0.0022	0.617	0.50	0.9662	0.9797
17	18.4	−1627.1	1.63	0.1045	0.0017	0.664	0.61	0.9663	0.9804
18	20.1	−1624.4	−0.60	0.5525	0.0012	0.744	0.64	0.9657	0.9804
19	17.4	−1626.5	2.18	0.0307	0.0013	0.462	0.78	0.9652	0.9813
20	19.0	−1623.8	−0.59	0.5539	0.0006	0.806	0.97	0.9656	0.9811

Table 1. Nine measures at stepwise selection of variables having maximal covariance.

No	(i)	(ii)	(iii)	(iv)	(v)	(vi)	(vii)	(viii)	(ix)
10	87.4	−1591.7	3.64	0.0004	0.0411	0.058	0.34	0.9668	0.9815
11	75.4	−1599.5	−3.24	0.0014	0.0419	0.033	0.33	0.9662	0.9814
12	60.2	−1610.5	−3.69	0.0003	0.0301	0.000	0.33	0.9645	0.9789
13	49.7	−1618.4	−3.24	0.0014	0.0052	0.622	0.36	0.9643	0.9798
14	33.8	−1632.0	4.01	0.0001	0.0088	0.028	0.86	0.9645	0.9807
15	30.3	−1634.3	−2.26	0.0249	0.0073	0.117	2.25	0.9663	0.9794
16	26.2	−1637.4	2.40	0.0175	0.0056	0.365	2.25	0.9677	0.9780
17	**23.8**	**−1638.9**	**−2.05**	**0.0422**	**0.0076**	**0.023**	**2.28**	**0.9701**	**0.9778**
18	22.7	−1639.1	1.76	0.0807	0.0026	0.541	2.38	0.9698	0.9789
19	20.2	−1640.9	2.10	0.0370	0.0011	0.821	3.33	0.9698	0.9823
20	19.0	−1641.3	1.79	0.0752	0.0021	0.541	4.72	0.9684	0.9836

Table 2. Nine measures at stepwise selection of variables having maximal increase in R^2.

No	(i)	(ii)	(iii)	(iv)	(v)	(vi)	(vii)	(viii)	(ix)
10	854.5	−1250.7	−2.40	0.0175	0.0586	0.000	0.15	0.7984	0.9176
11	452.2	−1348.6	−11.08	0.0000	0.0555	0.000	0.16	0.8722	0.9521
12	421.2	−1356.5	3.24	0.0014	0.0322	0.000	0.20	0.8783	0.9498
13	171.2	−1464.5	−11.72	0.0000	0.0302	0.000	0.19	0.9334	0.9682
14	172.8	−1461.7	−0.48	0.6294	0.0108	0.000	0.23	0.9292	0.9674
15	97.6	−1509.7	−7.32	0.0000	0.0108	0.000	0.27	0.9421	0.9718
16	66.2	−1533.4	5.12	0.0000	0.0071	0.000	0.32	0.9508	0.9729
17	**27.3**	**−1568.8**	**−6.21**	**0.0000**	**0.0053**	**0.000**	**0.35**	**0.9566**	**0.9729**
18	29.2	−1565.8	0.36	0.7170	0.0023	0.000	0.43	0.9563	0.9721
19	31.1	−1562.8	0.25	0.7991	0.0023	0.000	0.54	0.9553	0.9722
20	19.0	−1574.8	3.75	0.0002	0.0023	0.000	0.67	0.9579	0.9743

Table 3. Nine measures at principal component regression.

However, it may not always be a good practice. The *t*-test is not a valid test and the score vectors beyond dimension 20 are so small that they have no practical importance.

8.2.4. Ridge regression

Here score vectors are selected that correspond to the associated eigenvalues of $\mathbf{S} = \mathbf{X}^T\mathbf{X} + k\mathbf{I}$. The value of k is estimated by leave-one-out cross-validation, see reference [1]. The optimal value of k is $k = 0.0002$ (**Table 4**).

It seems also appropriate here to choose dimension 17. The value of k is based on the full model. However, there is no practical difference between dimension 17 and a full model.

8.2.5. PLS regression

When working with PLS regression, each step is analysed closer. The score vectors associated with the test set are computed as $\mathbf{T}_t = \mathbf{X}_t \times \mathbf{V}$. The correlation coefficient between response values of the test set, $\mathbf{Y}_t$, and the 12th score vector is –0.058 and the associated *p*-value is 0.197. Therefore, we can conclude that the 12th score vector does not contribute to the modelling task. We can thus conclude that the dimension should be 11 (**Table 5**).

8.2.6. H-method, maximal R^2 value

Here, it can be recommended to use 13 components. When 13 have been selected, there is no covariance left (see steps (v) and (vi)) (**Table 6**).

No	(i)	(ii)	(iii)	(iv)	(v)	(vi)	(vii)	(viii)	(ix)
10	782.7	–1250.6	–2.39	0.0176	0.0586	0.000	0.15	0.7984	0.9176
11	410.6	–1347.7	–1103	0.0000	0.0555	0.000	0.16	0.8722	0.9520
12	382.1	–1355.4	3.22	0.0015	0.0322	0.000	0.20	0.8783	0.9497
13	151.7	–1461.5	–11.59	0.0000	0.0302	0.000	0.19	0.9334	0.9681
14	153.3	–1458.6	–0.48	0.6345	0.0108	0.000	0.23	0.9292	0.9673
15	85.8	–1504.0	–7.10	0.0000	0.0108	0.000	0.28	0.9421	0.9717
16	57.9	–1525.8	4.93	0.0000	0.0071	0.000	0.32	0.9508	0.9731
17	**23.7**	**–1557.7**	**–5.90**	**0.0000**	**0.0053**	**0.000**	**0.35**	**0.9566**	**0.9734**
18	25.6	–1554.8	0.34	0.7345	0.0023	0.000	0.43	0.9563	0.9727
19	27.5	–1551.7	0.23	0.8168	0.0023	0.000	0.53	0.9553	0.9728
20	19.0	–1560.0	3.24	0.0014	0.0023	0.000	0.65	0.9579	0.9744

Table 4. Nine measures at ridge regression.

No	(i)	(ii)	(iii)	(iv)	(v)	(vi)	(vii)	(viii)	(ix)
10	207.0	–1514.5	7.14	0.000	0.0296	0.000	0.11	0.9426	0.9738
11	**137.0**	**–1553.0**	**6.59**	**0.000**	**0.0116**	**0.000**	**0.14**	**0.9498**	**0.9735**
12	107.8	–1570.9	4.55	0.000	0.0100	0.000	0.17	0.9568	0.9725
13	67.3	–1600.5	5.75	0.000	0.0101	0.000	0.20	0.9587	0.9744
14	58.0	–1607.1	3.02	0.003	0.0024	0.563	0.26	0.9628	0.9754
15	39.7	–1622.5	4.22	0.000	0.0040	0.000	0.35	0.9656	0.9762
16	26.6	–1634.4	3.78	0.000	0.0024	0.088	0.43	0.9654	0.9775
17	22.4	–1637.7	2.43	0.016	0.0023	0.013	0.47	0.9671	0.9804
18	23.5	–1635.6	0.96	0.340	0.0005	0.849	0.52	0.9673	0.9804
19	21.1	–1637.3	2.08	0.039	0.0004	0.822	0.85	0.9679	0.9826
20	19.0	–1638.7	2.02	0.045	0.0006	0.641	1.12	0.9674	0.9816

Table 5. Nine measures at PLS regression.

No	(i)	(ii)	(iii)	(iv)	(v)	(vi)	(vii)	(viii)	(ix)
10	243.3	–1488.5	8.86	0.000	0.0334	0.000	0.13	0.9499	0.9731
11	116.1	–1559.0	9.13	0.000	0.0168	0.000	0.18	0.9556	0.9759
12	75.4	–1587.5	5.65	0.000	0.0096	0.005	0.21	0.9576	0.9757
13	**68.3**	**–1591.9**	**2.65**	**0.009**	**0.0086**	**0.001**	**0.22**	**0.9592**	**0.9767**
14	38.1	–1617.4	5.33	0.000	0.0027	0.546	0.50	0.9615	0.9753
15	30.9	–1623.2	2.90	0.004	0.0040	0.020	0.56	0.9637	0.9764
16	26.5	–1626.6	2.45	0.015	0.0022	0.362	0.59	0.9654	0.9798
17	22.5	–1629.8	2.42	0.016	0.0018	0.275	0.63	0.9671	0.9823
18	20.8	–1630.7	1.91	0.058	0.0011	0.633	0.70	0.9674	0.9825
19	19.0	–1631.7	1.94	0.054	0.0008	0.691	0.81	09674	0.9815
20	19.0	–1630.8	1.41	0.161	0.0006	0.744	0.97	0.9672	0.9816

Table 6. Nine measures at H-method determining maximal R^2 value along the covariance.

	Stepw. max cov	**Stepw.** max R^2	**PCR**	**Ridge regression**	**PLS regression**	**H-method Max R^2**
$(1-\|y_n-\hat{y}_n\|^2/(y_n^T y_n))$	0.9346	0.9346	0.9330	0.9333	**0.9472**	0.9391
$\sqrt{Var}(\mathbf{b})$	0.46	2.28	0.35	0.35	**0.14**	0.22

Table 7. Comparison of results from six methods.

8.3. Evaluation of modelling results

When modelling has been carried out, industry standards recommend to apply the model to 40 new samples, $(\mathbf{X}_n, \mathbf{y}_n)$. The estimated values are $\hat{\mathbf{y}}_n = \mathbf{X}_n\mathbf{b}$. $\mathbf{X}_n$ are the new X-values and $\mathbf{b}$ is the regression coefficients from the method. Results are shown in **Table 7**.

From **Table 7**, it can be seen that PLS regression is slightly better than the other methods. The table also shows the estimate of the standard deviations of the regression coefficients. It shows that the PLS regression has much smaller value than the other methods. In conclusion, the PLS regression can be recommended for determining y-values for future samples.

Note that the values in the first row in **Table 7** are smaller than those were obtained for cross-validation and test sets during the calibration analysis. It indicates that the new 40 samples deviate in some way from the 160 samples that were used in the analysis. This is not explored further here.

8.4. Confidence interval for regression coefficients

Approximate confidence intervals for regression coefficients can be obtained by the procedure presented in reference [10]. Let e_i be the ith residual obtained by a regression method. Define

$$h_i = \mathbf{x}^i \mathbf{S}_A^{-1} \mathbf{x}^{iT} \text{ and } r_i = \frac{e_i}{1 - h_i}, \ i = 1, 2, \ldots, N. \tag{49}$$

A new set of residuals e_i^* are defined by randomly sampling from the modified residuals $r_1, \ldots, r_N$. By repeating the generation of (e_i^*), a new set of regression coefficients are obtained. This can be repeated say 200 times to get a confidence interval for the regression coefficients.

9. Discussion

Interpretation of **Tables 1–6** reflects personal experience. Others may select the dimension in a different way. The main issue is that the dimension of all the selected methods is less than 20. If a full model including all 40 variables is estimated, we are overfitting by a dimension over 20. Regression coefficients become large and inference from the estimation not reliable.

The advantage of working with latent variable is that the first ones collect a large amount of variation. This is well known from principal component analysis. Interpretation of variables cannot be done directly in latent variable models. However, using Eqs. (17) and (18) we can set up equations that show how individual variables contribute to the fit and regression coefficients.

10. Applications of the H-principle

The H-principle has been extended to many areas of applied mathematics. It has in general been successful due to the presence of latent structure in data. The implementation of the H-principle is different from area to area. Furthermore, 'household' administration may be needed in order to keep the number of variables low (cf. Mallows theory). In many cases, it has opened up for new mathematics.

In reference [11], it has been extended to multi-block and path modelling, giving new methods to carry out modelling of organized data blocks. The importance of these methods is due to the fact that the methods of regression analysis are extended so that regression models are computed between data blocks. Methods of regression analysis can be used to evaluate relationships between data blocks. Complicated assumptions like we see in structural equations modelling are not needed.

Linear latent structure regression can be viewed as determining a low-dimensional hyperplane in a high-dimensional space. In reference [12], it is extended to finding low-dimensional second, third and higher order surfaces in latent variables. Deviations from linearity often appear as curvature for low and high sample values, which can be handled by these surfaces. In reference [13], it has been applied to non-linear estimation that may give good low-rank solutions, where full-rank regularized solutions do not give convergence.

In reference [14], it is extended to multi-linear algebra, where there are many indices in data, e.g. $\mathbf{X} = (x_{ijk})$ and $\mathbf{y} = (y_{ij})$. The basic issue in multi-linear algebra is defining the inverse. This is solved by defining directional inverses for each dimension. It makes it possible to extend methods of ordinary matrix analysis to multi-linear algebra. These methods have been successfully applied to multi-linear data and to growth models.

The H-principle has been applied to several areas of applied statistics. Here we briefly mention a few.

In time series analysis and dynamic systems, the objective of modelling is both to describe the data and also to obtain good forecasts. Thus, the requirement to a latent variable is both that it describes **X** and also gives good forecasts. Traditional models only focus on the description of **X**. By requiring that the latent variables also should give good forecast, better models are obtained.

In pattern recognition and classification, the objective of modelling is both to obtain good description of each group of data (that are detected or given a-priori) and to get low-error rate of classification. There are different ways to implement the H-principle in these areas, which have been developed. Applications show that these methods are superior to those based on statistical theory (e.g. linear and quadratic discriminate analysis based on the normal distribution, principal component analysis of each group of data).

11. Conclusion

We have presented a short review of standard regression analysis. Theoretically, it has important properties, which makes it a standard approach in popular program packages. However, we show that the results obtained may not always be reliable, when there is a latent structure in data. This is a serious problem in industry and applied sciences, because data typically have latent structure.

The H-principle is formulated in close analogy to the Heisenberg uncertainty inequality. It suggests that in the case of uncertain data, the computation of the mathematical solution should be carried out in steps, where at each step an optimal balance between the fit and associated precision should be obtained. A general framework is presented for linear regression. Any set of weight vectors can be used that do not give loading vectors of zero size. Using the framework, six different regression methods are carried out. It is shown that the methods give low rank solutions. The same type of numerical and graphic analysis can be carried out for any type of regression analysis within this framework. Traditional analysis in applied statistics and graphic analysis, which is popular in chemometrics, can be carried out for each method that uses the framework.

The algorithm can be viewed as an approximation to the full rank solution. Modelling stops, if further steps are not supported by data. Dimension measures, cross-validation and test sets are used to identify, when the steps are supported by data.

Acknowledgements

The cooperation with Clinical Biochemistry, Holbæk Hospital, Denmark, in the Sime project is highly appreciated. The author appreciates the use of the data from the project in the present article.

Author details

Agnar Höskuldsson

Address all correspondence to: ah@agnarh.dk

Centre for Advanced Data Analysis, Denmark

References

[1] Höskuldsson A, A common framework for linear regression, *Chemometrics and Intelligent Laboratory Systems* 146 (2015) 250–262. DOI: 10.1016/j.chemolab.2015.05.022

[2] Reinikainen SP, Höskuldsson A, COVPROC method: strategy in modeling dynamic systems, *Journal of Chemometrics* 17 (2003) 130–139. DOI: 10.1002/cem.770

[3] McLeod G, et al. A comparison of variate pre-selection methods for use in partial least squares regression: a case study on NIR spectroscopy applied to monitoring beer fermentation, *Journal of Food Engineering* 90 (2009) 300–307. DOI: 10.1016/j.jfoodeng.2008.06.037

[4] Tapp HS, et al., Evaluation of multiple variate methods from a biological perspective: a nutrigenomics case study, *Genes Nutrition* 7 (2012) 387–397. DOI: 10.1007/s12263-012-0288-4

[5] *https://en.wikipedia.org/wiki/**Mallows's_Cp***

[6] *https://en.wikipedia.org/wiki/**Akaike_information_criterion***

[7] Siotani M, Hayakawa T, Fujikoshi Y, *Modern Multivariate Analysis: A Graduate Course and Handbook*. American Science Press: Columbus, Ohio, 1985.

[8] Höskuldsson A: *Prediction Methods in Science and Technology*, Vol. 1, Thor Publishing: Copenhagen, 1996, ISBN 87-985941-0-9

[9] Clinical and Laboratory Standards Institute, *http://shop.clsi.org/chemistry-documents/*

[10] Davison AC, Hinkley DV: *Bootstrap Methods and their Application*, Cambridge University Press: Cambridge, New York, 1997.

[11] Höskuldsson A, Modelling procedures for directed network of data blocks, *Chemometrics and Intelligent Laboratory Systems* 97 (2009) 3–10. DOI: 10.1016/j.chemolab.2008.09.002

[12] Höskuldsson, A. The Heisenberg modelling procedure and applications to nonlinear modelling. *Chemometrics and Intelligent Laboratory System* 44 (1998) 15–30. DOI: 10.1016/S0169-7439(98)00111-7

[13] Höskuldsson A, H-methods in applied sciences, *Journal of Chemometrics* 22 (2008) 150–177. DOI: 10.1002/cem.1131

[14] Höskuldsson A, Data analysis, matrix decompositions and generalized inverse, *SIAM Journal on Scientific Computing*, 15 (1994) 239–262. DOI: 10.1137/0915018

2

A Statistic Method for Anatomical and Evolutionary Analysis

Roqueline Ametila e Gloria Martins de Freitas Aversi-Ferreira,
Hisao Nishijo and Tales Alexandre Aversi-Ferreira

Abstract

Rules, formulas, and statistical tests have been widely used in studies that analyze continuous variables with the normal (Gaussian) distribution or defined parameters. Nevertheless, in some studies such as those in gross anatomy, only statistics with discrete or nominal variables are available. In fact, the existence or absence of an anatomical structure, its features and internal aspects, innervation, arterial and vein supplies, etc. can be analyzed as discrete and/or nominal variables. However, there have been no adequate methods, which allow transformation of data with qualitative/nominal variables in gross anatomy to those with quantitative variables. To resolve the issue, we have purposed a new method that allows, in order, descriptions based on numerical analyses, the statistical method for comparative anatomy (SMCA), and proposed the formula for comparison of groups of anatomical structures among different species that allows to infer evolutionary perspective. The important features of this method are as follows: (1) to allow to analyze numerical data, which are converted from discrete or nominal variables in morphological areas and (2) to quantitatively compare identical structures within the same species and across different species. The SMCA fills the lack of a specific method for statistical works in comparative anatomy, morphology, in general, and evolutional correlations.

Keywords: statistic nonparametric, nominal variables, anatomy, morphology, evolution

1. Introduction

The statistical analysis is widely used in studies in almost all scientific fields to lead to discussions and conclusions of the data [1]. In most of the cases, the variables analyzed are continuous ones that can be analyzed by calculating an average, standard deviation, i.e., the data can be fitted approximately to Gaussian distribution (normal distribution) of probability. Gaussian distribution means that most population, in studied variables, is concentrated around the average, i.e., the data are grouped around the average symmetrically [2]; these data, when fitted in Cartesian plane, display a geometrical shape similar to an inverted bell, called Gaussian curve showing a central tendency. In a perfect Gaussian distribution, the average is located at the center of the curve and the frequency of the data, in a studied variable, decreases quantitatively toward lateral extremes. This type of data distribution, similar distribution around the average, is called parametric statistic.

It is important that data characteristics regarding their distribution must be analyzed before any statistical calculation. However, variables of data that do not follow Gaussian distribution are sometimes submitted to statistical calculation under assumption of normal distribution of probability. In these cases, application to mathematical tools under assumption of normal distribution would induce errors, to be more specific, acceptance or rejection of statistical hypotheses should be incorrect [1]. These kinds of errors are becoming common mainly because of indiscriminate and incorrect use of statistical software. The statistical programs are very important to allow fast, concise, and reliable analyses of variables and include a test of normality of data. However, sometimes, data that do not display normal distribution are analyzed using these programs, and the results usually indicate no statistical significance.

For understanding the importance of correct statistics, imagine that although average atmospheric temperature around a man is 24°C, his feet are put in a refrigerator and the head is put in a stove. Then, he should consider the temperatures as very uncomfortable [3]. In this case, the average hardly reflects correct interpretation of the data. Indeed, these extreme values could provide an acceptable average. However, mostly these values are not fitted into Gaussian distribution. Therefore, they cannot be analyzed by parametric statistics [1–3].

Data that cannot be submitted to parametric designs should be analyzed using nonparametric statistics. Indeed, other types of averages can be calculated based on nonnormal distributions, as in the binomial or chi-square (χ^2) distribution, for example. However, they usually do not allow central tendency analyses and require randomness among data. Furthermore, nonparametric statistics are less precise than parametric ones [4].

In gross anatomy, there are many cases in which numerical data are not available for analyses. Anatomical studies must analyze absence or presence of a structure or organ and characteristics associated with these organs; for example, presence or absence of specific nerves and vessels in muscles, and distribution of these structures if these structures are present. They are qualitative variables, but not numerical ones. This means that numbers cannot provide this type of information.

It is well-known that anatomical texts include vast descriptions of structures, relationships between the structures, axes, and positions of the body. These findings indicate that specific statistical methods are required especially in comparative anatomical studies. However, any previous statistical methods do not allow accurate analyses of anatomical data, but only could assist discussion on them. Some anatomical studies tried to analyze qualitative variables more objectively, using nonparametric statistics such as chi-square (χ^2) [5]. The basis of the chi-square statistics is associativity among the data. Thus, the chi-square statistic is an important tool as a multivariate analysis of discrete variables that are considered to be independent. However, the statistical hypothesis in this test does not agree with the Darwin's theory of evolution, *inter alia*, because of the assumption of existence of a common ancestor [6]. Indeed, the assumption of the common ancestor suggests similarities of structures across species, i.e., they cannot be random in organisms that evolved from a same ancestor, since the ancestral animal provided basic structures, and could provide derivative features with descendants (for a detailed review, see Ref. [7]). Therefore, application to the statistical methods such as chi-square statistics, which are based on the randomized premise, could induce a hypothetical error.

It is reasonable that when the central tendency measures cannot be used, nonparametric distributions must be chosen [2], mainly due to small sample sizes [3]. The nonparametric distributions are also used when it is difficult to set up quantitative variables. Indeed, the percentage of a given structure based on the frequency of the structure in the samples is one of the key measures in nonparametric statistics used in gross anatomy. In gross anatomy, the highest percentage of occurrence of a given structure is called **normal**, while the lowest percentage is called **variation** [8]. Percentages of **normal** and **variation** could be analyzed using nonparametric methods.

Gross anatomy has no specific statistical method for analyses of noncontinuous variables regarding anatomical structures until a few years ago. Here, we show a new statistical method based on nonparametric statistics, more consistent with anatomical descriptions. We also compare this new method with cladistics used for evolutionary analyses to indicate usefulness of this new method in this discipline.

2. Concepts of the statistical methods for gross anatomy

In this section, we will show that the new statistical method is based on the anatomical concept of normality, and appropriate weight is provided with each variable (parameter for a specific feature) of structures based on the importance of the variable, and that conclusions can be drawn based on the values integrated across multiple variables. The results by this new method have been reported in our previous papers, in which this method was designed to compare muscles not only within the same species but also across different species in comparative gross anatomy [1, 7, 9–11].

2.1. Anatomical concept of normality and variation

The initial step in the statistical method for comparative anatomy (**SMCA**) is to calculate the frequency based on the **normality** and **variation** concept in anatomy. "A normal structure" means that it is observed in greater than 50% of cases within the same species. Therefore, the variation can be observed in less than 50% of cases [8]. The summary of the steps to calculate SMCA is shown in **Table 1**.

Good examples of structures in animals that should be applied to SMCA are muscles, because muscles require different variables to describe their characteristics: shapes, innervation, vascularization, origin, insertion, and number. Different individuals in the same species or individuals in different species could display different numbers, as in the contrahentes muscles in primates.

The formula indicating the relationship among a total number of studied structures and numbers of normal and variation in the structures is shown below:

$$N = N_{.j.} = \sum_{i=1}^{q} \left(r_{v(ijk)} + n_{v(ijk)} \right) \quad (1)$$

where $\boldsymbol{N}$ is the total number of analyzed structures, $\boldsymbol{n_v}$ is the number of structures with variation, and $\boldsymbol{r_v}$ is the number of normal structures $(\boldsymbol{N\text{-}n_v})$. The subscript i indicates specific species, for instance human, chimpanzees, etc., while the subscript j indicates specific structures, and the subscript k indicates parameters (variables) of the specific structures. Thus, the sum of normal and variation in structures must be a total (100%) of these structures.

In case of muscles, the parameters should include at least the following four: (1) innervation, (2) origin, (3) insertion, and (4) vascularization. For instance, in a case of the biceps in a specific species, j is 1 (j = 1), and i is 1 (i = 1). The data analysis in this step should be performed in terms of the following four parameters: (1) innervation ($r_{v(111)}$, $n_{v(111)}$), (2) origin ($r_{v(112)}$, $n_{v(112)}$), (3) insertion ($r_{v(113)}$, $n_{v(113)}$), and (4) vascularization ($r_{v(114)}$, $n_{v(114)}$). Furthermore, (5) number of muscles ($r_{v(115)}$, $n_{v(115)}$) and (6) shape ($r_{v(116)}$, $n_{v(116)}$) could be added for more detailed analyses. In addition, further detailed parameters (subscript (h)) could be added (see below in detail).

The next step is calculation of the relative frequency (**RF** = $\boldsymbol{P_{ijk}}$) of normal structures in each parameter against the total number of structures based on frequencies of normality and variation, i.e., (1) innervation ($r_{v(111)}$, $n_{v(111)}$), (2) origin ($r_{v(112)}$, $n_{v(112)}$), (3) insertion ($r_{v(113)}$, $n_{v(113)}$), and (4) vascularization ($r_{v(114)}$, $n_{v(114)}$). According to these frequencies, each RF for (1) innervation ($\boldsymbol{P_{ij1}}$), (2) origin ($\boldsymbol{P_{ij2}}$), (3) insertion ($\boldsymbol{P_{ij3}}$), and (4) vascularization ($\boldsymbol{P_{ij4}}$) can be calculated, as follows:

$$\mathrm{RF} = P_{ijk} = \frac{r_{v(ijk)}}{N} \quad (2)$$

When the structure is pair organs, **N** (number of individuals in a sample) must be multiplied by 2. It is also possible to separately calculate $\boldsymbol{P_{ijk}}$ for each piece in each body side in case of pair organs. Although any values can be used for **N**, smaller number of **N** will result in lower statistical power. It is obviously essential to analyze large numbers of specimens. The analyses with small numbers of specimens are not appropriate for scientific analyses.

It is noted that qualitative features are transformed into quantitative data after the initial data are expressed as percentages. Thus, the method allows numerical description of anatomical structures, which increases preciseness in description of characteristics of anatomical structures. Another usefulness of this method is that the value of **RF** can be obtained from previous literatures as prevalence (percentage of the structure) in a given species. This is especially impor-

Species ($i = 1, 2,\ldots, s$)		**Specie 1 (control specie) ($i = 1$)**				**Specie 2 ($i = 2$)**				…	**Specie s ($i = s$)**			
Investigated structures ($j = 1, 2,\ldots, m$)		M_1	M_2	…	M_m	M_1	M_2	…	M_m	…	M_1	M_2	…	M_m
P_{ijk}	Weights given													
Innervation ($k = 1$)	$w_1(=3)$	P_{111}	P_{121}	…	P_{1m1}	P_{211}	P_{221}	…	P_{2m1}	…	P_{s11}	P_{s21}	…	P_{sm1}
Origin ($k = 2$)	$w_2(=2)$	P_{112}	P_{122}	…	P_{1m2}	P_{212}	P_{222}	…	P_{2m2}	…	P_{s12}	P_{s22}	…	P_{sm2}
Insertion ($k = 3$)	$w_3(=2)$	P_{113}	P_{123}	…	P_{1m3}	P_{213}	P_{223}	…	P_{2m3}	…	P_{s13}	P_{s23}	…	P_{sm3}
Vascularization ($k = 4$)	$w_4(=1)$	P_{114}	P_{124}	…	P_{1m4}	P_{214}	P_{224}	…	P_{2m4}	…	P_{s14}	P_{s24}	…	P_{sm4}
Weighted averages for single muscle (PAF = $P_{w(ij)}$)		$P_{w(11)}$	$P_{w(12)}$	…	$P_{w(1m)}$	$P_{w(21)}$	$P_{w(22)}$	…	$P_{w(2m)}$	…	$P_{w(s1)}$	$P_{w(s2)}$	…	$P_{w(sm)}$
Weighted averages for multiple muscles (mean of $P_{w(ij)}$) ($P_{w(i)}$)		$P_{w(1)}$				$P_{w(2)}$				…	$P_{w(s)}$			
GCAI= $\lvert P_{w(i)} - P_{w(i')}\rvert$						$\lvert P_{w(1)} - P_{w(2)}\rvert$				…	$\lvert P_{w(1)} - P_{w(s)}\rvert$			

Note: P_{ijk} = proportion of normal structured organs, where i represents individual species, j represents individual structures (i.e., muscles), and k represents individual parameters of the muscles.

Table 1. Illustration of steps to compute GCAI based on multiple structures in different species having muscles as example, according to Aversi-Ferreira et al.

tant when the study includes comparative anatomy. For example, the palmaris longus could be defect in humans [8, 10, 12] and its prevalence is around 90% [13]; therefore, **RF** might be 90% among total individuals. However, in the analyses of innervation, vascularization, origin, or insertion of the palmaris longus, only 90% receives attention and the data from the remaining 10% are sometimes discarded. Such case is common in comparative studies, where, usually, only data in specific species are studied.

Normal structure in each parameter means $0.5 < \boldsymbol{P}_{ijk} \leq 1$ in practical terms, because a quantity lesser than 50% does not match the definition of normal structures in anatomy. However, mathematically, $\boldsymbol{P}_{ijk}$ can vary as follows: $0 \leq \boldsymbol{P}_{ijk} \leq 1$. For a given species, $\boldsymbol{P}_{ijk}$, according to the concept of normality, must be greater than 0.5. However, when different species are compared, the frequencies of the **normal** structures could be different. For example, when the dorsoepitrochlearis muscle is compared among primates and *Homo*, it is rarely observed in modern humans [14] and approximate $\boldsymbol{P}_{ijk}$ is 0.05 (this value was derived from the literature reporting the percentage of presence of this muscle in individuals), while, in nonhuman primates, the dorsoepitroclearis muscle is a normal feature, and the $\boldsymbol{P}_{ijk}$ is 1.00. Furthermore, some muscles have more than one origin or insertion, as in the triceps brachii with three heads, and in rare cases, this muscle has four heads of origin in *Homo* [8, 13]. Therefore, there are only two kinds of origins in this muscle; type 1 containing three heads as a normal feature and type 2 containing four heads as a variation form.

For accurate and detailed analyses, it is required to calculate $\boldsymbol{P}_{ijk}$ by adding other multiple parameters for muscles or other structures. For example, for muscles, the parameters should include, at least, (1) number or kinds of nerves, or branches of a same nerve ($\boldsymbol{P}_{ij1}$), (2) origin(s) of muscles ($\boldsymbol{P}_{ij2}$), (3) insertion of muscles ($\boldsymbol{P}_{ij3}$), and (4) vascularization of muscles by arteries or branches of one artery ($\boldsymbol{P}_{ij4}$). These parameters should be chosen according to the goal of the analysis; some of the parameters could be removed while the other could be added. Furthermore, (5) quantity of muscles ($\boldsymbol{P}_{ij5}$) and its (6) shape ($\boldsymbol{P}_{ij6}$) could be included in more detailed studies. It is noted that small number of parameters results in less characterization of the studied structure.

By the introduction of this parameter ($\boldsymbol{P}_{ijk}$) in this method (SMCA), anatomical characteristics (shown by $\boldsymbol{P}_{ijk}$) can be compared among different samples within the same species or across different species, which is useful characteristic of the SMCA for studies of comparative anatomy. For example, variations in the number of the muscles could be compared within the same species or across different species in primates [12].

2.2. Definition of pondered average of frequency (PAF)

In the next step of the SMCS in which multiple features ($\boldsymbol{P}_{ijk}$) of a given structure are compared among different species, a unique variable (PAF), an integrated value over multiple parameters ($\boldsymbol{P}_{ijk}$) is computed. For this purpose, pondered values [the weighted coefficients (w_k)], multiplied by $\boldsymbol{P}_{ijk}$, are specified. The coefficients must be specified according to the anatomical importance of a given parameter in assessment of anatomical similarity. For example, since a small value of the $\boldsymbol{P}_{ijk}$ is ascribed to large variations, the characteristic is not important in assessment of structure similarity. Therefore, the P_{ijk} with small values must be associated with small weighted coefficients. On the other hand, the P_{ijk} with large values (i.e., few variations) must be associated with larger weighted coefficients (see below).

After designation of pondered values as weighted coefficients, the pondered average of frequencies (**PAF** = $P_{w(ij)}$) is computed according to the following formula:

$$\mathbf{PAF} = P_{w(ij)} = \frac{\sum_{k=1}^{q} w_k \cdot P_{ijk}}{\sum_{k=1}^{q} w_k};$$

for any species ($i = 1, 2, \ldots, s$) and any muscles ($j = 1, 2, \ldots, m$); (3)

where P_{ijk} is the relative frequency and w_k is the weighted coefficient linked to a specific parameter. For instance, for the muscle 1 of species 1, P_{111} is the PAF (relative frequency) of innervation, and weighted coefficient w_1 is 3; P_{112} is PAF of the muscle origin, and w_2 is 2; P_{113} is relative frequency of muscle insertion, and w_3 is 2; and P_{114} is relative frequency of vascularization, and w_4 is 1 [9].

The idea of weighted coefficients (w_k) is based on frequencies of variation in studied structures. The structures with less variation could receive larger weight and the more variable structures, smaller weight. For example, if vessels receive larger weight in the comparison, this could compromise final results, leading to an idea of larger anatomical difference among specimens or species in spite of less difference in muscles. Therefore, we gave the weighted coefficient 3 to **innervation** ($k = 1$, $w_1 = 3$) in a case of muscles. In muscles, during embryonic development of animals, a given nerve terminates on a given muscle [14]. Thus, variations in innervation of muscles are few. Therefore, variations in the innervation is very sensitive to the differences among individuals in a same species and, also, to the difference among different species. Accordingly, the four parameters for muscles noted above, i.e., innervation, origin, insertion, and vascularization, the innervation shows less variations, **origin** and **insertion** usually show similar variations, and vascularization, more variations. Thus, both weighted coefficients for origin and insertion should receive the same weight coefficient 2 ($w_2 = 2$ for origin, $w_3 = 2$ for insertion). Finally, the parameter with greater variation, **vascularization** ($k = 4$), received the weighted coefficient 1 ($w_4 = 1$). Indeed, vascularization can be different between the same muscles in bilateral sides within the same individuals [12].

Zero cannot be accepted as weighted coefficient (w_k). Therefore, w_k must be greater than zero, i.e., $w_k > 0$. To make the calculation easier and to keep clear parameters, the best choice is to use only integer values, i.e., $w_k \geq 1$. Accordingly, it is very important to keep in mind the choice of the weighted coefficients should depend on different degrees of variations of the structures; highest weighted coefficient for the parameter with the lowest variations, or the same weighted coefficients for the parameters with identical degree of variation. The designed numbers also should be integers or discrete ones since it does not make sense to look for values that represent the exact differences among descriptive or nominal variables.

2.3. Definition of comparative anatomy index (CAI) for comparison among different species

In normal structures, $P_{w(ij)}$ must be greater than 0.5 and less than or equal to 1, i.e., $0.5 < P_{w(ij)} \leq 1$. In fact, $P_{w(ij)}$ could be 1, if every $P_{w(ij)}$ has maximal value 1, and if every $P_{w(ij)}$ is minimum ($P_{w(ij)} > 0.5$), the $P_{w(ij)}$ will be 0.5, as well. Mathematically, $P_{w(ij)}$ can vary within the range of $0 \leq P_{w(ij)} \leq 1$, since $P_{w(ij)}$ could be zero or less than 0.5 in analyses using different species in which such structure might not be normal.

It is noted that $P_{w(ij)}$ [which is a function of each relative frequency of each specific feature (P_{ijk}) and its weight (w_k)] can be used to assess mathematical similarity of a population of anatomical data; equal values indicate high similarity and large difference in the values between two species indicates dissimilarities or less similarity. In order to compare structures among different species, $P_{w(ij)}$ has to be calculated in each species.

Before calculating $P_{w(ij)}$, each P_{ijk} must be computed according to the data of species used as a reference, i.e., the **control species**. For example, the coracobrachialis, a muscle of the arm, could have one or two cranial heads in different primates [15]. For the coracobrachialis, P_{ijk} could be different depending on the number of cranial heads in the **control species**. Thus, P_{ijk} must be consistently calculated in reference to **control species**, since different species could have different normal structures (see below in detail).

For example, there are two types of origin in the coracobrachialis ($j = 1$); type 1 has one origin and type 2 has two origins ($k = 2$). P_{ijk} could take different values according to the number of heads in the reference species (**control species**) ($i = 1$). In noncontrol species to be studied ($i = 2$) in which type 1 (number of origin is 1) is normal, the P_{212} will be 1 in reference to the species with one head, and P_{212} of type 1 will be 0.5 in reference to the species in which type 2 (two heads) is normal. In case of the muscle that has one to three heads of origins in different species, the P_{ijk} should be divided by maximum number (i.e., 3) of heads resulting in 1/3, because P_{ijk} should not be greater than 1. Accordingly, when **control species** with three heads of cranial origin of a muscle is chosen as reference, then, P_{212} is 1.000 (3 heads, $i = 2$), P_{312} is 0.667 (two heads, $i = 3$), and P_{412} is 0.333 (1 head, $i = 4$).

Therefore, P_{ijk} must be obtained initially for the **control species**. When the **control species** ($i = 1$) have anatomically normal two cranial heads ($k = 2$) for the coracobrachialis ($j = 1$), and if all individuals in this species have two cranial heads (100%), P_{112} should be 1; if 90% of individuals in this species have two cranial heads, P_{112} should be 9/10. In a case of a noncontrol species ($i = 2$) in which the anatomical normal is one cranial head for the coracobrachial, if all individuals have one head, the P_{212} should be 0.5, and if 90% of the individuals have one head, P_{212} should be 0.45. These values noted above (P_{ijk}) can be obtained from the data in previous studies, and can be applied to the CAI analysis (see below).

Although any species can be defined as control species, the species studied in the first time or the species with abundant known data should be chosen as control species. To compare any single structure (e.g., muscle) between two different species ($i \neq i'$), the data in any noncontrol species can be compared one by one with those in the control species using the comparative anatomy index (CAI) defined by the following the formula:

$$\mathbf{CAI} = \left| P_{w(ij)} - P_{w(i'j')} \right| \text{ ; where } i \neq i' \quad (4)$$

The $\mathbf{CAI}_{ii'}$ formula represents the absolute difference of weighted averages ($P_{w(ij)}$) of a single structure between the **control** *i* and other noncontrol *i′* species. In this comparison, the non-control *i′* species should be always compared with the same **control species** *i*. For example, the formula to compare a structure ($j = 1$) with a given feature (parameter) ($k = 1$) between the ***i*** and ***i′*** species is shown as follows:

$$\mathbf{CAI} = \left| P_{w(11)} - P_{w(21)} \right|. \quad (5)$$

It is noted that the $\mathbf{CAI}_{ii'}$ ranges from 0 to 1, i.e., $0 \leq \mathbf{CAI}_{ii'} \leq 1$. This is because the maximum value of $P_{w(ij)}$ is 1 and the minimum is 0. Note that this equation permits only comparison of just one structure between the two species.

2.4. Definition of group comparative anatomy index (GCAI) for comparison of a group of structures among different species

However, the **SMCA** analysis of the muscles in the forearm [9] shows the need to compare many muscles of a same functional group between different species, for instance, the deep flexor muscles in the forearm among studied species of primates. This need could be understood as reasonable purpose because they work together for some functions as close the hand, then, the comparison of these muscles as a group seems more appropriate in terms of physiology, phylogeny, taxonomy, and evolution, as well. Thus, was purposed the **GCAI** to compare a group of the muscles among species [1, 9, 11], one by one based on the sum of the $P_{w(ij)}$, as follows:

$$P_{w(i)} = \frac{\sum_{j=1}^{m_j} P_{w(ij)}}{m_j} \quad (6)$$

where ***i*** indicates number of species ($i = 1, 2, \ldots, s$) and ***j*** indicates number of studied structures ($j = 1, 2, \ldots, m$) and m_j is the number of structures studied in a sample. Usually, m_j is m ($m_j = m$) because, usually, equal quantity of structures is studied in each species.

The **GCAI**, which represents difference in $P_{w(i)}$ based on multiple structures between the **control (*i*)** and other noncontrol (***i′***) species, is defined by the following formula:

$$\mathbf{GCAI}_{ii'} = \left| \mathrm{P}_{w(i)} - \mathrm{P}_{w(i')} \right| \quad (7)$$

or

$$\mathbf{GCAI}_{ii'} = \frac{\sum_{j=1}^{m_j} P_{w(ij)}}{m_j} - \frac{\sum_{j'=1}^{m_{j'}} P_{w(ij')}}{m_{j'}}; \quad (8)$$

Based on the above inferences, using **SMCA**, the values close to 0.000 suggest **high similarity** of the structures between the species, and the value 1.000 indicates that those are completely different structures. Thus, we can rank similarity among species based on SMCA. For example, we can define the CAI or GCAI values of 0 as high similarity among the structures analysed, the values from 0 to 0.200 as similar structures, those from 0.200 to 0.650 as somewhat similar, and those from 0.650 to 1.000 as dissimilar. Thus, the **GCAI** is the absolute difference in mean weighted averages of $P_{w(ij)}$ for multiple muscles between the two species.

In **Table 2**, we show that different structures (muscles) among different species can be compared in reference to a control species. The 8 specimens of the *Sapajus* sp, 16 forearms with a total of the 304 muscles, were analyzed [9], and the data derived from the previous literatures using chimpanzees, gorillas, baboons, and humans were compared with the muscles in the control species (*Sapajus* sp). For other examples of SMCA application to muscles, see Aversi-Ferreira et al. [7, 9–11]. In the same way, we also analyzed 4 Japanese monkeys for 4 muscles in the arm resulting in a total of the 16 structures [11], which were compared with the data of the modern humans and other primates obtained from previous studies except those for *Sapajus* sp.

Muscle	Features	*Bearded capuchin (BC)* [chosen as control species]	*Men*	*Chimpanzee*	*Baboon*
		Superficial dorsal group of the forearm			
Extensor digitorum communis	Origin	Lateral epicondyle of the humerus	Lesser variation regarding the distribution of tendons to fingers. Somewhat similar to BC CAI = 0.22	Highly similar to BC CAI = 0.00	Highly similar to BC CAI = 0.00
	Insertion	Dorsal aponeurosis in the second to fifth proximal phalanges			
	Innervation	Radial nerve			
	Vascularization	Radial artery			
Extensor digiti quinti proprius	Origin	Lateral epicondyle of the humerus	Only one insertion tendon to little finger. Somewhat similar to BC CAI = 0.22	Fleshy portion is well detached. Somewhat similar to BC CAI = 0.22	Highly similar to BC CAI = 0.0
			GCAI = 0.22; somewhat similar to BC	GCAI = 0.11; similar to BC	GCAI = 0.00; highly similar to BC

Note: Observe that the simple description is company by a numerical analysis, i.e., qualitative features were transformed in a numerical data for objective to compare different structures of different species. The values of CAI or GCAI chosen for comparison effect were of 0 as high similarity among the structures analysed, from 0 to 0.200 as similar structures, from 0.200 to 0.650 as somewhat similar, from 0.650 to 1.000 as dissimilar.

Table 2. Examples of the analysis using SMCA applied to muscles of the forearm.

For humans, variation of the structures is very well informed, and too for great apes. However, others animals, except for domestic ones, are scarcely studied, and if any the number of the specimens or species could be small. However, calculation of the SMCA based on the previous studies, except the values of the *N*, will provides quantitative data for analyses of morphological distances of structures among the species or those within the same species.

2.5. Comparison of the SMCA with other nonparametric statistics

Another possibility to study nominal variables is to use the cladistics method that is commonly used in evolutionary studies. This method supposes binary characteristic of data, and any other possibility could be an error, because these features are mutually exclusives [15, 16]. This method is useful to obtain objective and/or precise information of evolution-related structures regarding the absence or presence of such structure across different species. However, this characteristic limits its application to morphological analyses of structures, since it considers just two parameters; 0 for absent characteristic, and 1 for its presence. Nevertheless, this method is important in evolutionary studies, since this method might provide evolutionary information. Cladistics analyses prioritize the primitive and derivative features [15, 16], while the morphological analyses studied here (SMCA) prioritize utmost characters observed in a given structure.

We previously compared SMCA with other nonparametric methods including cladistics [7, 10]. In fact, the SMCA accept more variables to be analyzed for each structure than the cladistics method. For a more detailed comparison, see Aversi-Ferreira et al. [7].

3. Conclusion

It is desirable to quantitatively assess any kinds of data, even in gross anatomy [7, 10], which is important for more precise discussions and more reliable conclusions [17]. Indeed, according to Lord Kelvin, "When you can measure what you are speaking about, and express it in numbers, you know something about it." Our objective is to provide a statistical test for gross anatomy to numerically compare structures of different subjects within the same species and those across different species, which should be useful to analyse more precisely and objectively the data in comparative anatomy.

The SMCA is a new statistical method and requires further verification using many data. We reported SMCA analyses previously [7, 9–11] and the SMCA could satisfactorily incorporate many qualitative data numerically. In conclusion, the main features of SMCA are as follows: (1) to allow numerical description of the data shown by discrete or nominal variables in comparative anatomy or in other areas of morphology and (2) to provide a, at least, more precise (numerical) method for comparison of samples of structures from the same species and from different species. Thus, the SMCA fills the lack of an appropriate method for statistical works in comparative anatomy, and in other areas of morphology and other disciplines such as taxonomy, phylogenetic, and evolution.

Acknowledgments

T.A. Aversi-Ferreira is a recipient of Scholarship Research Productivity from National Council of Technology and Development (CNPq/Brazil). This work was supported partly by the Japan Society for the Promotion of Science (JSPS), Grant-in-Aid for Scientific Research (B) (16H04652). The authors declare no competing financial interests.

Author details

Roqueline Ametila e Gloria Martins de Freitas Aversi-Ferreira[2], Hisao Nishijo[1] and Tales Alexandre Aversi-Ferreira[1,3,*]

*Address all correspondence to: aversiferreira@gmail.com

1 System Emotional Science, Department of Physiology, School of Medicine and Pharmaceutical Sciences, University of Toyama, Toyama, Japan

2 Nursing and Pharmacy School, FAPAL, Tocantins, Brazil

3 Institute of Biomedical Sciences, University of Alfenas, MG, Brazil

References

[1] Aversi-Ferreira TA. A new statistical method for comparative anatomy. Int. J. Morphol. 2009; 27: 1051–1058.

[2] Vieira S. Biostatistics: advanced topics. 2nd ed. Rio de Janeiro: Elsevier; 2004. 216p.

[3] Centeno AJ. Statistical course applied to biology. 2nd ed. Goiânia: Editora da Universidade de Goiás; 2001. 235p.

[4] Kitchen CR. Nonparametric versus parametric tests of location in biomedical research. Am J Ophthalmol. 2009; **147**: 571–572. DOI:10.1016/j.ajo.2008.06.

[5] Barros RAC, Prada ILS, Silva Z, Ribeiro AR, Silva DCO. Lumbar plexus formation of the *cebus apella* monkey. Brazil J Veter Res Anim Sci. 2003; **40**: 373–381. DOI:10.1590/S1413-95962003000500009

[6] Darwin CR. Mutual affinities of organic beings: morphology: embryology: rudimentary organs. In: The Origin of Species. 6th ed. London: John Murray; 1859. pp. 411–458.

[7] Aversi-Ferreira RAGMF, Nishijo H, Aversi-Ferreira TA. Reexamination of statistical methods for comparative anatomy: examples of its application and comparisons with other parametric and nonparametric statistics. BioMed Res Int. 2015, Article ID 902534. doi:10.1155/2015/902534

[8] Moore KL, Dalley AF, Agur AMR. Clinically oriented anatomy. 7th ed. Philadelphia: Wolters Kluwer Health/Lippincott Williams & Wilkins; 2014. 1173p.

[9] Aversi-Ferreira TA, Maior RS, Carneiro-e-Silva FO, Aversi-Ferreira RAGMF, Tavares MC, Nishijo H, Tomaz C. Comparative anatomical analyses of the forearm muscles of *cebus libidinosus* (Rylands *et al.* 2000): Manipulatory behavior and tool use. PLoS One. 2011; **6**: e22165. DOI:10.1371/journal.pone.0022165

[10] Aversi-Ferreira RAGMF, Bretas RV, Rafael Souto Maior RS, Davaasuren M, Paraguassu-Chaves CA, Nishijo H, Aversi-Ferreira TA. Morphometric and statistical analysis of the palmaris longus muscle in human and non-human primates. BioMed Res Int. 2014; Article ID 178906. DOI:10.1155/2014/178906

[11] Aversi-Ferreira TA, Aversi-Ferreira RA, Bretas RV, Nishimaru H, Nishijo H. Comparative anatomy of the arm muscles of the Japanese monkey (*macaca fuscata*) with some comments on locomotor mechanics and behavior. J Med Primatol. 2016; **45**:165–79. DOI: 10.1111/jmp.12222.

[12] Swindler DR, Wood CD. An Atlas of Primates Gross Anatomy: Baboon, Chimpanzee and Man. Seattle: University of Washington Press; 1973. 370p.

[13] Standring S, editor. Pelvis girdle and lower limb. In: Gray's anatomy: The Anatomical Basis of Clinical Practice. London: Churchill Livingstone; 2008. pp. 1327–1464.

[14] Gray H. Miology. In: Anatomy of the Human Body. 1918. Available from: http://www.bartleby.com/107. [Accessed: 2016-08-15]

[15] Pimentcl RA, Riggins R. The nature of cladistic data. Cladisticis. 1987; **3**: 201–209. DOI: 10.1111/j.1096-0031.1987.tb00508.x

[16] Hawkins A, Hughes CE, Scotland bRW. Primary homology assessment, characters and character states. Cladistics. 1997; **13**:275–283. DOI: 10.1111/j.1096-0031.1997.tb00320.x

[17] Thomson W. Electrical units of measurement. In: The Practical Applications of Electricity. London: The Institution of Civil Engineers; 1884. pp. 149–176.

3

Why the Decision-Theoretic Perspective Misrepresents Frequentist Inference: Revisiting Stein's Paradox and Admissibility

Aris Spanos

Abstract

The primary objective of this paper is to make a case that R.A. Fisher's objections to the decision-theoretic framing of frequentist inference are not without merit. It is argued that this framing is congruent with the Bayesian but incongruent with the frequentist approach; it provides the former with a theory of optimal inference but misrepresents the optimality theory of the latter. Decision-theoretic and Bayesian rules are considered optimal when they minimize the expected loss "for all possible values of θ in Θ" [$\forall\theta\in\Theta$], irrespective of what the true value θ^* [state of Nature] happens to be; the value that gave rise to the data. In contrast, the theory of optimal frequentist inference is framed entirely in terms of the capacity of the procedure to pinpoint θ^*. The inappropriateness of the quantifier $\forall\theta\in\Theta$ calls into question the relevance of admissibility as a minimal property for frequentist estimators. As a result, the pertinence of Stein's paradox, as it relates to the capacity of frequentist estimators to pinpoint θ^*, needs to be reassessed. The paper also contrasts loss-based errors with traditional frequentist errors, arguing that the former are attached to θ, but the latter to the inference procedure itself.

Keywords: decision theoretic inference, Bayesian vs. frequentist inference, Stein's paradox, James-Stein estimator, loss functions, admissibility, error probabilities, loss functions, risk functions, complete class theorem

1. Introduction

Wald's [1] decision-theoretic framework is widely viewed as providing a broad enough perspective to accommodate and compare the frequentist and Bayesian approaches to inference, despite their well-known differences. It is perceived as offering a neutral framing of inference that brings into focus their common features and tones down their differences; see Refs. [2–4].

Historically, Wald [5] proposed the original variant of the decision-theoretic framework with a view to unify Neyman's [6] rendering of frequentist interval estimation and testing:

"The problem in this formulation is very general. It contains the problems of testing hypotheses and of statistical estimation treated in the literature." (p. 340)

Among the frequentist pioneers, Jerzy Neyman accepted enthusiastically this broader perspective, primarily because the concepts of decision rules and action spaces seemed to provide a better framing for his *behavioristic interpretation* of Neyman-Pearson (N-P) testing based on the accept/reject rules; see Refs. [7, 8]. Neyman's attitude towards Wald's [1] framing was also adopted wholeheartedly by some of his most influential students/colleagues at Berkeley, including [9, 10]. In a foreword of a collection of Neyman's early papers, his students/editors described the Wald's framing as ([11], p. vii):

"A natural but far reaching extension of their [N-P formulation] scope can be found in Abraham Wald's theory of statistical decision functions."

At the other end of the argument, Fisher [12] rejected Wald's framing on the grounds that it seriously distorts his rendering of frequentist statistics:

"The attempt to reinterpret the common tests of significance used in scientific research as though they constituted some kind of acceptance procedure and led to "decisions" in Wald's sense, originated in several misapprehensions and has led, apparently, to several more." (p. 69)

With a few exceptions, such as Refs. [13–15], Fisher's [12] viewpoint has been inadequately discussed and evaluated by the subsequent statistics literature. The primary aim of this paper is to revisit Fisher's minority view by taking a closer look at the decision-theoretic framework with a view to reevaluate the claim that it provides a neutral framework for comparing the frequentist and Bayesian approaches. It is argued that Fisher's view that the decision theoretic framing is germane to "acceptance sampling," but misrepresents frequentist inference, is not without merit. The key argument of the discussion that follows is that the decision-theoretic notions of loss function and admissibility are congruent with the Bayesian approach, but incongruent with both the primary objective and the underlying reasoning of the frequentist approach.

Section 2 introduces the basic elements of the decision theoretic set-up with a view to bring out its links to the Bayesian and frequentist approaches, calling into question the conventional wisdom concerning its neutrality. Section 3 takes a closer look at the Bayesian approach and argues that had the decision-theoretic apparatus not exist, Bayesians would have been forced to invent it in order to establish a theory of optimal Bayesian inference. Section 4 discusses critically the notions of loss functions and admissibility, focusing primarily on their role in giving rise to Stein's paradox and their incompatibility with the frequentist approach. It is argued that the frequentist dimension of the notions of a loss function and admissibility is more apparent than real. Section 5 makes a case that the decision-theoretic framework misrepresents both the primary objective and the underlying reasoning of the frequentist approach. Section 6 revisits the notion of a loss function and its dependence on "information other than the data." It is argued that loss-based errors are both different and incompatible with the traditional frequentist errors because they are attached to the unknown parameters instead of the inference procedures themselves, as the traditional frequentist errors (Type I, II and coverage).

2. The decision theoretic set-up

2.1. Basic elements of the decision-theoretic framing

The current decision-theoretic set-up has three basic elements:

1. A prespecified (parametric) statistical model $\mathcal{M}_\theta(\mathbf{x})$, generically specified by

$$\mathcal{M}_\theta(\mathbf{x})=\{f(\mathbf{x};\theta),\ \theta\in\Theta\},\ \mathbf{x}\in\mathbb{R}^n_X,\ \text{for}\ \ \theta\in\Theta\subset\mathbb{R}^m,\ m\ll n, \tag{1}$$

where $f(\mathbf{x};\theta)$ denotes the (joint) *distribution of the sample* $\mathbf{X}:=(X_1,\ldots,X_n)$, $\mathbb{R}^n_X$ denotes the *sample space* and Θ the *parameter space*. This model represents the stochastic mechanism assumed to have given rise to data $\mathbf{x}_0:=(x_1,\ldots,x_n)$.

2. A decision space D containing all mappings $d(.):\ \mathbb{R}^n_X\to A$, where A denotes the set of all actions available to the statistician.

3. A loss function $L(.,.):\ [D\times\Theta]\to R$, representing the numerical loss if the statistician takes action $a\in A$ when the state of Nature is $\theta\in\Theta$; see Refs. [2, 16–18].

The basic idea is that when the decision-maker selects action a, he/she does not know the "true" state of Nature, represented by θ^*. However, contingent on each action $a\in A$, the decision maker "knows" the losses (gains and utilities) resulting from different choices $(d,\theta)\in[D\times\Theta]$. The decision maker observes data $\mathbf{x}_0$, which provides some information about θ^* and then maps each $\mathbf{x}\in\mathbb{R}^n_X$ to a certain action $a\in A$ guided solely by $L(d,\theta)$.

2.2. The original Wald framing

It is important to bring out the fact that the original Wald [5] framing was much narrower than the above basic elements 2 and 3, due to its original objective to formalize the Neyman-Pearson (N-P) approach; see [19]. What were the key differences?

i. The decision (action) space D was defined exclusively in terms of subsets of the parameter space Θ. For estimation purposes $D:=\{\theta:\theta\in\Theta\}$ is the set of all singleton points of Θ and for testing $D:=(\Theta_0,\Theta_1)$, the null and alternative regions, respectively.

ii. The original loss (weight) was a zero-positive function, with zero loss at:

$$L_{0-c}(\theta,\widehat{\theta}(\mathbf{X}))=\begin{cases}0\ \text{if}\ \widehat{\theta}(\mathbf{X})=\theta^*\\ c_\theta>0\ \text{if}\ \widehat{\theta}(\mathbf{X})=\theta\neq\theta^*,\ \theta\in\Theta,\end{cases} \tag{2}$$

where θ^* is the true value of θ in Θ. For the discussion that follows, it is important to note that Eq. (2) is nonoperational in practice because θ^* is unknown.

The more general framing, introduced by Wald ([1, 20]) and broadened by Le Cam [21], extended the scope of the original set-up by generalizing the notions of loss functions and decision spaces. In what follows it is argued that these extensions created serious incompatibilities with both the objective and the underlying reasoning of frequentist inference.

In addition, it is both of historical and methodological interest to note that Wald [5] introduced the notion of a prior distribution, $\pi(\theta)$, $\forall\theta\in\Theta$, into the original decision-theoretic machinery reluctantly, and justified it on being a useful tool for proving certain technical results:

"The situation regarding the introduction of an a priori probability distribution of θ is entirely different. First, the objection can be made against it, as Neyman has pointed out, that θ is merely an unknown constant and not a variate, hence it makes no sense to speak of the probability distribution of θ. Second, even if we may assume that θ is a variate, we have in general no possibility of determining the distribution of θ and any assumptions regarding this distribution are of hypothetical character. The reason why we introduce here a hypothetical probability distribution of θ is simply that it proves to be useful in deducing certain theorems and in the calculation of the best system of regions of acceptance." (p. 302)

2.3. A shared neutral framework?

The frequentist, Bayesian, and the decision-theoretic approaches share the notion of a statistical model by viewing data $\mathbf{x}_0 := (x_1, \ldots, x_n)$ as a realization of a sample $\mathbf{X} := (X_1, \ldots, X_n)$ from Eq. (1).

The key differences between the three approaches are as follows:

a. The frequentist approach relies exclusively on $\mathcal{M}_\theta(\mathbf{x})$

b. The Bayesian approach adds a *prior* distribution, $\pi(\theta)$, $\forall\theta\in\Theta$ (for all $\theta\in\Theta$)

c. The decision-theoretic framing revolves around a *loss* (gain or utility) function:

$$L(d(\mathbf{x}),\theta),\forall\theta\in\Theta,\ \forall\mathbf{x}\in\mathbb{R}^n_X. \tag{3}$$

The loss function is often assumed to be an even, differentiable and convex function of $(d(\mathbf{x})-\theta)$ and can take numerous functional forms; see Refs. [17, 18] inter alia.

The claim that the decision-theoretic perspective provides a neutral ground is often justified [3] on account of the loss function being a function of the sample and parameter spaces through the two *universal quantifiers:*

(i) "$\forall\mathbf{x}\in\mathbb{R}^n_X$," associated with the *distribution of the sample*:

$$\textbf{frequentist}:\ \ f(\mathbf{x};\theta),\ \forall\mathbf{x}\in\mathbb{R}^n_X, \tag{4}$$

(ii)"$\forall\theta\in\Theta$" associated with the *posterior distribution*:

$$\textbf{Bayesian}:\ \ \pi(\theta|\mathbf{x}_0)=\frac{\pi(\theta)\cdot f(\mathbf{x}_0|\theta)}{\int_{\theta\in\Theta}\pi(\theta)\cdot f(\mathbf{x}_0|\theta)d\theta},\ \forall\theta\in\Theta. \tag{5}$$

The idea is that allowing for all values of $\mathbf{x}$ in $\mathbb{R}^n_X$ goes beyond the Bayesian perspective, which relies exclusively on a single point $\mathbf{x}_0$. What is not obvious is whether that is sufficient to do justice to the frequentist approach. A closer scrutiny suggests that frequentist inference is misrepresented by the way both quantifiers are employed in the decision-theoretic framing of inference.

First, the quantifier $\forall \mathbf{x} \in \mathbb{R}^n_X$ plays only a minor role in transforming a loss function, say $L(\theta,\hat{\theta}(\mathbf{x}))$, into a risk function:

$$R(\theta,\hat{\theta})=E_{\mathbf{X}}[L(\theta,\hat{\theta}(\mathbf{X}))]=\int_{\mathbf{x}\in\mathbb{R}^n_X} L(\theta,\hat{\theta}(\mathbf{x}))f(\mathbf{x};\theta)d\mathbf{x}, \forall\theta\in\Theta. \tag{6}$$

This is the only place where the distribution of the sample, $f(\mathbf{x};\theta)$, $\forall \mathbf{x} \in \mathbb{R}^n_X$ enters the decision-theoretic framing, and the only relevant part of the behavior of $\hat{\theta}(\mathbf{X})$ is how it affects the risk function for different values of θ in Θ. In frequentist inference, however, the distribution of the sample takes center stage for the theory of optimal frequentist inference. It determines the sampling distribution of any statistic $Y_n=g(\mathbf{X})$ (estimator, test, and predictor) through:

$$F(y;\theta) := \mathbb{P}(Y_n \leq y;\theta) = \underbrace{\int\int\cdots\int}_{\{\mathbf{x}:\, g(\mathbf{x})\leq t;\ \mathbf{x}\in\mathbb{R}^n_X\}} f(\mathbf{x};\theta)d\mathbf{x}, \tag{7}$$

and that, in turn, yields the relevant error probabilities that determine optimal inference procedures.

Second, the decision-theoretic notion of optimality revolves around the universal quantifier "$\forall\theta\in\Theta$," rendering it congruent with the Bayesian but incongruent with the frequentist approach. To be more specific, since different risk functions often intersect over Θ, an optimal rule is usually selected after the risk function is reduced to a scalar. Two such choices of risk are:

$$\begin{aligned} &\textbf{Maximum risk}: \quad R_{\max}(\hat{\theta})=\sup_{\theta\in\Theta} R(\theta,\hat{\theta}),\\ &\textbf{Bayes risk}: \quad R_B(\hat{\theta})=\int_{\theta\in\Theta} R(\theta,\hat{\theta})\pi(\theta)d\theta. \end{aligned} \tag{8}$$

Hence, an obvious way to choose among different rules is to find the one that minimizes the relevant risk with respect to all possible estimates $\tilde{\theta}(\mathbf{x})$. In the case of Eq. (8), this gives rise to two corresponding decision rules:

$$\begin{aligned} &\textbf{Minimax rule}: \quad \inf_{\theta(\mathbf{x})} R_{\max}(\hat{\theta})=\inf_{\theta(\mathbf{x})}[\sup_{\theta\in\Theta} R(\theta,\hat{\theta})],\\ &\textbf{Bayes rule}: \inf_{\tilde{\theta}(\mathbf{x})} R_B(\hat{\theta})=\inf_{\tilde{\theta}(\mathbf{x})}\int_{\theta\in\Theta} R(\theta,\hat{\theta})\pi(\theta)d\theta. \end{aligned} \tag{9}$$

In this sense, a decision or a Bayes rule $\tilde{\theta}(\mathbf{x})$ will be considered optimal when it minimizes the relevant risk, no matter what the true state of Nature θ^* happens to be. The last clause, "irrespective of θ^*" constitutes a crucial caveat that is often ignored in discussions of these approaches. When viewed as a game against Nature, the decision maker selects action a from A, irrespective of what value θ^* Nature has chosen. That is, θ^* plays no role in selecting the optimal rules since the latter have nothing to do with the true value θ^* of θ. To avoid any misreading of this line of reasoning, it is important to emphasize that "the true value θ^*" is shorthand for saying that "data $\mathbf{x}_0$ constitute a typical realization of the sample $\mathbf{X}$ with distribution $f(\mathbf{x};\theta^*)$"; see Ref. [22].

This should be contrasted with the notion of optimality in frequentist inference that gives θ^* center stage, in the sense that it evaluates the capacity of the inference procedure to inform the modeler about θ^*; no other value is relevant. According to Reid [23]:

"A statistical model is a family of probability distributions $[\mathcal{M}_\theta(\mathbf{x})]$, the central problem of statistical inference being to identify which member of the family $[\theta^*]$ generated the data of interest." (p. 418)

3. The Bayesian approach

To shed further light on the affinity between the decision-theoretic framework and the Bayesian approach, let us take a closer look at the latter.

3.1. Bayesian inference and its primary objective

A key argument in favor of the Bayesian approach is often its simplicity in the sense that all forms of inference revolve around a single function, the posterior distribution: $\pi(\theta|\mathbf{x}_0) \propto \pi(\theta) \cdot f(\mathbf{x}_0|\theta),\ \forall\theta\in\Theta$. Hence, an outsider looking at Bayesian approach might naturally surmise that its primary objective is to yield "a probabilistic ranking" (ordering) of all values of θ in Θ. According to O'Hagan [4]:

"Having obtained the posterior density $\pi(\theta|\mathbf{x}_0)$, the final step of the Bayesian method is to derive from it suitable inference statements. The most usual inference question is this: After seeing the data $\mathbf{x}_0$, what do we now know about the parameter θ. The only answer to this question is to present the entire posterior distribution." (p. 6)

The idea is that the modeling begins with an a priori probabilistic ranking based on $\pi(\theta),\ \forall\theta\in\Theta$, which is revised after observing $\mathbf{x}_0$ to derive $\pi(\theta|\mathbf{x}_0),\ \forall\theta\in\Theta$; hence the key role of the quantifier $\forall\theta\in\Theta$. O'Hagan [4], echoing earlier views in [24, 25], contrast the frequentist (classical) inferences with the Bayesian inference arguing:

"Classical inference theory is very concerned with constructing *good* inference rules. The primary concern of Bayesian inference, …, is entirely different. The objective is to extract information concerning θ from the posterior distribution, and to present it helpfully via effective summaries. There are two criteria in this process. The first is to identify interesting features of the posterior distribution. … The second criterion is good communication. Summaries should be chosen to convey clearly and succinctly all the features of interest. … In Bayesian terms, therefore, a good inference is one which contributes effectively to appropriating the information about θ which is conveyed by the posterior distribution." (p. 14)

Clearly, O'Hagan's [4] attempt to define what is a "good" Bayesian inference begs the question: what does constitute "effective appropriation of information about θ" mean, beyond the probabilistic ranking? That is, the issue of optimality is inextricably bound up with what the primary objective of Bayesian inference is. If the primary objective of Bayesian inference is not the revised probabilistic ranking, what is it? The answer is that the ranking is only half the story. The other half is concerned with the optimality for Bayesian inference which cannot be framed exclusively in terms of the posterior distribution. The decision-theoretic perspective provides the Bayesian approach with a theory of optimal inference as well as a primary objective: minimize expected losses for all values of θ in Θ.

In his attempt to defend his stance that the entire posterior distribution *is* the inference, O'Hagan [4] argues that criteria for "optimal" Bayesian inferences are only *parasitical* on the Bayesian approach and enter the picture through the decision theoretic perspective:

"… a study of decision theory has two potential benefits. First, it provides a link to classical inference. It thereby shows to what extent classical estimators, confidence intervals and hypotheses tests can be given a Bayesian interpretation or motivation. Second, it helps identify suitable summaries to give Bayesian answers to stylized inference questions which classical theory addresses." (p. 14)

Both of the above mentioned potential benefits to the Bayesian approach, are questionable for two reasons. First, the link between the decision-theoretic and the classical (frequentist) inference is more apparent than real because it is fraught with misleading definitions and unclarities pertaining to the reasoning and objectives of the latter. As argued in the sequel, the quantifier "$\forall\theta\in\Theta$" used to define "optimal" decision-theoretic or Bayes rules is at odds with and misrepresents frequentist inference. Second, the claim concerning Bayesian answers to frequentist questions of interest is misplaced because the former provides no real answers to the frequentist primary question of interest which pertains to learning about θ^*. An optimal Bayes rule offers very little, if anything, relevant for learning about the value θ^* that gave rise to $\mathbf{x}_0$. Let us unpack this answer in some more detail.

3.2. Optimality for Bayesian inference

What does minimizing the Bayes risk amount to? Substituting the risk function in Eq. (6) into the Bayes risk in Eq. (8), one can show that:

$$\begin{aligned} R_B(\hat{\theta}) &= \int_{\theta\in\Theta}\left(\int_{\mathbf{x}\in\mathbb{R}^n_X} L(\theta,\hat{\theta}(\mathbf{x}))f(\mathbf{x};\theta)d\mathbf{x}\right)\pi(\theta)d\theta \\ &= \int_{\mathbf{x}\in\mathbb{R}^n_X}\int_{\theta\in\Theta} L(\theta,\theta(\mathbf{x}))f(\mathbf{x}|\theta)\pi(\theta)d\theta d\mathbf{x} \\ &= \int_{\mathbf{x}\in\mathbb{R}^n_X}\left\{\int_{\theta\in\Theta} L(\theta,\theta(\mathbf{x}))\pi(\theta|\mathbf{x})d\theta\right\}m(\mathbf{x})d\mathbf{x}, \end{aligned} \tag{10}$$

where $m(\mathbf{x})=\int_{\theta\in\Theta} f(\mathbf{x};\theta)d\theta$; see Ref. [18]. The second and third equalities presume that one can reverse the order of integration (a technical issue), and treat $f(\mathbf{x};\theta)$ as the joint distribution of $\mathbf{X}$ and θ so that the following equalities hold:

$$f(\mathbf{x};\theta)=f(\mathbf{x}|\theta)\pi(\theta)=\pi(\theta|\mathbf{x})m(\mathbf{x}). \tag{11}$$

In this case, these equalities are questionable due to the blurring of the distinction between $\mathbf{x}$, a generic value of $\mathbb{R}^n_X$, and the particular value $\mathbf{x}_0$; see Ref. [26].

In light of Eq. (10), a Bayesian estimate is "optimal" relative to a particular loss function $L(\hat{\theta}(\mathbf{X}),\theta)$, when it minimizes $R_B(\hat{\theta})$, or equivalently $\int_{\theta\in\Theta} L(\theta,\hat{\theta}(\mathbf{x}))\pi(\theta|\mathbf{x})d\theta$. This makes it clear that what constitutes an "optimal" Bayesian estimate is primarily determined by $L(\hat{\theta}(\mathbf{X}),\theta)$ [27]:

i. When $L_2(\hat{\theta},\theta)=(\hat{\theta}-\theta)^2$, the Bayes estimate $\hat{\theta}$ is the *mean* of $\pi(\theta|\mathbf{x}_0)$.

ii. When $L_1(\tilde{\theta},\theta)=|\tilde{\theta}-\theta|$, the Bayes estimate $\hat{\theta}$ is the *median* of $\pi(\theta|\mathbf{x}_0)$.

iii. When $L_{0-1}(\overline{\theta},\theta)=\delta(\overline{\theta},\theta)=\begin{cases}0 & \text{for}|\overline{\theta}-\theta|<\varepsilon\\ 1 & \text{for}|\overline{\theta}-\theta|\geq\varepsilon\end{cases}$, for $\varepsilon>0$, the Bayes estimate $\overline{\theta}$ is the *mode* of $\pi(\theta|\mathbf{x}_0)$.

In practice, the most widely used loss function is the square:

$$L_2(\hat{\theta}(\mathbf{X});\theta)=(\hat{\theta}(\mathbf{X})-\theta)^2,\ \forall\theta\in\Theta, \tag{12}$$

whose risk function is the decision-theoretic *Mean Square Error* (MSE_1):

$$R(\theta,\hat{\theta})=E(\hat{\theta}(\mathbf{X})-\theta)^2=\text{MSE}_1(\hat{\theta}(\mathbf{X});\theta),\ \forall\theta\in\Theta. \tag{13}$$

Surprising, however, this definition of the MSE, denoted by MSE_1, is different from the frequentist MSE, which is defined by:

$$MSE(\hat{\theta}_n(\mathbf{X});\theta^*)=E(\hat{\theta}_n(\mathbf{X})-\theta^*)^2. \tag{14}$$

The key difference is that Eq. (14) is defined at the point $\theta=\theta^*$, as opposed to $\forall\theta\in\Theta$. Unfortunately, statistics textbooks adopt one of the two definitions of the MSE— either at $\theta=\theta^*$ or $\forall\theta\in\Theta$—and ignore (or seem unaware) of the other. At first sight, his difference might appear pedantic, but it turns out that it has very serious implications for the relevant theory of optimality for the frequentist vs. Bayesian inference procedures. Indeed, reliance on $\forall\theta\in\Theta$ undermines completely the relevance of admissibility as a minimal property for estimators in frequentist inference.

Admissibility. An estimator $\tilde{\theta}(\mathbf{X})$ is *inadmissible* if there exists another estimator $\hat{\theta}(\mathbf{X})$ such that:

$$R(\theta,\hat{\theta})\leq R(\theta,\tilde{\theta}),\forall\theta\in\Theta, \tag{15}$$

and the strict inequality (<) holds for at least one value of θ. Otherwise, $\tilde{\theta}(\mathbf{X})$ is said to be *admissible* with respect to the loss function $L(\theta,\hat{\theta})$.

The objective of minimizing losses weighted by $\pi(\theta|\mathbf{x}_0)$ for all value of θ in Θ, is in direct contrast to the frequentist primary objective, which is to learn from data about the true value θ^* underlying the generation of $\mathbf{x}_0$. Hence, the question that naturally arises is: what does an optimal Bayes rule, stemming from Eq. (17) convey about the underlying data generating mechanism in Eq. (1)? It is not obvious why the highest ranked value $\tilde{\theta}(\mathbf{x}_0)$ (mode), or some other feature of the posterior distribution, has any value in pinpointing θ^* knowing that $\tilde{\theta}(\mathbf{x}_0)$ is selected irrespective of θ^* the true state of Nature.

3.3. The duality between loss functions and priors

The derivation in Eq. (10) brings out the built-in affinity between the decision-theoretic framing of inference and the Bayesian approach. As shown above, minimizing the Bayes risk:

$$R_B(\hat{\theta}) = \int_{\theta\in\Theta} R(\hat{\theta},\theta)\pi(\theta)d\theta, \tag{16}$$

is equivalent to minimizing the integral:

$$\int_{\theta\in\Theta} L(\hat{\theta}(\mathbf{X}),\theta)\pi(\theta|\mathbf{x})d\theta. \tag{17}$$

This result brings out two important features of optimal Bayesian inference.

First, it confirms the minor role played by the quantifier $\mathbf{x}\in\mathbb{R}_X^n$ in both the Bayesian and decision-theoretic optimality theory of inference.

Second, it indicates that $L(\theta,\hat{\theta})$ and $\pi(\theta)$ are perfect substitutes with respect to any weight function $w(\theta) > 0$, $\forall\theta\in\Theta$, in the derivation of Bayes rules. Modifying the loss function or the prior yields the same result:

"... the problem of estimating θ with a modified (weighted) loss function is identical to the problem with a simple loss but with modified hyperparameters of the prior distribution while the form of the prior distribution does not change." ([28], p. 522)

This implies that in practice a Bayesian could derive a particular Bayes rule by attaching the weight to the loss function or to the prior distribution depending on which derivation is easier; see Refs. [18, 28].

3.4. Revisiting the complete class theorem

The issue of contrasting objectives highlights the key built-in tension between the frequentist and Bayesian approaches to optimality, which in turn undermines several important results, including the complete class theorem, first proved in Ref. [20]:

"Wald showed that under fairly general conditions the class of Bayes decision functions forms an essentially complete class; in other words, for any decision function that is not Bayesian, there exists one that is Bayes and is at least as good no matter what the true state of Nature may be." ([19], p. 341)

As argued in the sequel, it should come as no surprise to learn that Bayes rules dominate all other rules when admissibility is given center stage. The key result is that a Bayes rule $\hat{\theta}_B(\mathbf{x})$ with respect to a prior distribution $\pi(\theta)$ is:

i. *Admissible,* under certain regularity conditions, including when $\hat{\theta}_B(\mathbf{x})$ is unique up to equivalence relative to the same risk function $R(\theta,\hat{\theta}_B)$.

ii. *Minimax* when $R(\theta,\hat{\theta}_B) = c < \infty$.

iii. An admissible, relative to a risk function $R(\theta,\hat{\theta}_B)$, estimate $\hat{\theta}(\mathbf{x})$ is either Bayes $\hat{\theta}_B(\mathbf{x})$ or the limit of a sequence of Bayes rules; see Refs. [2, 17, 28].

Ignoring the contrasting objectives, these results have been interpreted as evidence for the superiority of the Bayesian perspective, and led to the intimation that an effective way to generate optimal frequentist procedures is to find the Bayes solution using a reasonable prior and then examine their frequentist properties to see whether it is satisfactory from the latter viewpoint; see Refs. [29, 30].

As argued next, even if one were to agree that Bayes rules and admissible estimators largely coincide, the importance of such a result hinges on the relevance of admissibility as a key property for frequentist estimators.

4. Loss functions and admissibility revisited

The claim to be discussed in this section is that the notions of a "loss function" and "admissibility" are incompatible with the optimal theory of frequentist estimation as framed by Fisher; see Ref. [31].

4.1. Admissibility as a minimal property

The following example brings out the inappropriateness of admissibility as a minimal property for optimal frequentist estimators.

Example. In the context of the simple Normal model:

$$X_k \sim \text{NIID}(\theta, 1),\ k=1,2,\ldots,n,\ \text{for}\ n>2, \tag{18}$$

consider the decision-theoretic notion of MSE_1 in Eq. (13) to compare two estimators of θ:

i. The maximum likelihood estimator (MLE): $\overline{X}_n=\frac{1}{n}\sum_{k=1}^{n} X_k$

ii. The "crystalball" estimator: $\theta_{cb}=7405926,\ \forall \mathbf{x}\in\mathbb{R}_X^n$

When compared on admissibility grounds, both estimators are admissible and thus equally acceptable. Common sense, however, suggests that if a particular criterion of optimality cannot distinguish between $\overline{X}_n$ [a strongly consistent, unbiased, fully efficient and sufficient estimator] and θ_{cb}, an arbitrarily chosen real number that ignores the data altogether, is not much of a minimal property.

A moment's reflection suggests that the inappropriateness of admissibility stems from its reliance on the quantifier "$\forall\theta\in\Theta$." The admissibility of θ_{cb} arises from the fact that for certain values of θ close enough to θ_{cb}, say $\theta \in (\theta_{cb} \pm \frac{\lambda}{\sqrt{n}})$, for $0 < \lambda < 1$, θ_{cb} is "better" than $\overline{X}_n$ on MSE_1 grounds:

$$MSE_1\left(\overline{X}_n; \theta\right) = \frac{1}{n} > MSE_1(\theta_{cb}; \theta) \leq \frac{\lambda^2}{n} \text{for } \theta \in \left(\theta_{cb} \pm \frac{\lambda}{\sqrt{n}}\right). \tag{19}$$

Given that the primary objective of a frequentist estimator is to pin-point θ^*, the result in Eq. (19) seems totally irrelevant as a gauge of its capacity to achieve that!

This example indicates that admissibility is totally ineffective as a *minimal* property because it does not filter out θ_{cb}, the worst possible estimator! Instead, it excludes potentially good estimators like the *sample median;* see Ref. [32]. This highlights the "extreme relativism" of admissibility to the particular loss function, $L_2(\hat{\theta}(\mathbf{X}); \theta)$, in this case. For the absolute loss function $L_1(\hat{\theta}(\mathbf{X}); \theta) = |\hat{\theta}(\mathbf{X}) - \theta|$, however, the sample median would have been the optimal estimator. Despite his wholehearted embrace of the decision-theoretic framing, Lehmann [33] warned statisticians about the perils of arbitrary loss functions:

"It is argued that the choice of a loss function, while less crucial than that of the model, exerts an important influence on the nature of the solution of a statistical decision problem, and that an arbitrary choice such as squared error may be baldly misleading as to the relative desirability of the competing procedures." (p. 425)

A strong case can be made that *the* key minimal property (necessary but not sufficient) for frequentist estimation is *consistency,* an extension of the Law of Large Numbers (LLN) to estimators, more generally. For instance, consistency would have eliminated θ_{cb} from consideration because it is inconsistent. This makes intuitive sense because if an estimator $\hat{\theta}(\mathbf{X})$ cannot pinpoint θ^* with an infinite data information, it should be considered irrelevant for learning about θ^*. Indeed, there is nothing in the notion of admissibility that advances learning from data about θ^*.

Further to relative (to particular loss functions) efficiency being a dubious property for frequentist estimators, the pertinent measure of finite sample precision for frequentist estimators is full efficiency, which is defined relative to the assumed statistical model (1).

4.2. Stein's paradox and admissibility

The quintessential example that has bolstered the appeal of the Bayesian claims concerning admissibility is the James-Stein estimator [34], which gave rise to an extensive literature on *shrinkage estimators,* see Ref. [35].

Let $\mathbf{X} := (X_1, X_2, \ldots, X_m)$ be independent sample from a Normal distribution:

$$X_k \sim \text{NI}(\theta_k, \sigma^2),\ k=1, 2, \ldots, m, \tag{20}$$

where σ^2 is known. Using the notation $\boldsymbol{\theta} := (\theta_1, \theta_2, \ldots, \theta_m)$ and $\mathbf{I}_m := \text{diag}(1, 1, \ldots, 1)$, this can be denoted by:

$$\mathbf{X}\sim\mathrm{N}(\theta,\ \sigma^2\mathbf{I}_m).$$

Find an optimal estimator $\tilde{\theta}(\mathbf{X})$ of θ with respect to the square "overall" loss function:

$$L_2(\boldsymbol{\theta},\hat{\boldsymbol{\theta}}(\mathbf{X}))=(\|\hat{\boldsymbol{\theta}}(\mathbf{X})-\boldsymbol{\theta}\|^2)=\sum_{k=1}^{m}(\hat{\theta}_k(\mathbf{X})-\theta_k)^2. \tag{21}$$

Stein [36] astounded the statistical world by showing that for $m=2$ the least-squares (LS) estimator $\hat{\boldsymbol{\theta}}_{LS}(\mathbf{X})=\mathbf{X}$ is admissible, but for $m>2$ $\hat{\boldsymbol{\theta}}_{LS}(\mathbf{X})$ is inadmissible. Indeed, James and Stein [37] were able to come up with a nonlinear estimator:

$$\hat{\boldsymbol{\theta}}_{JS}(\mathbf{X})=\left(1-\frac{(m-2)\sigma^2}{\|\mathbf{X}\|^2}\right)\mathbf{X}, \tag{22}$$

that became known as the James-Stein estimator, which dominates $\hat{\boldsymbol{\theta}}_{LS}(\mathbf{X})=\mathbf{X}$ in MSE_1 terms by demonstrating that:

$$\mathrm{MSE}_1(\hat{\boldsymbol{\theta}}_{JS}(\mathbf{X});\theta)<\mathrm{MSE}_1(\hat{\boldsymbol{\theta}}_{LS}(\mathbf{X});\theta),\forall\theta\in\mathbb{R}^m. \tag{23}$$

It turns out that $\hat{\boldsymbol{\theta}}_{JM}(\mathbf{X})$ is also inadmissible for $m>2$ and dominated by the modified James-Stein estimator that is *admissible*:

$$\hat{\theta}^{+}_{JS}(\mathbf{X})=\left(1-\tfrac{(m-2)\sigma^2}{\|\mathbf{X}\|^2}\right)^{+}\mathbf{X}, \tag{24}$$

where $(z)^+=\max(0,z);$ see Ref. [17].

The traditional interpretation of this result is that for the Normal, Independent model in Eq. (20), the James-Stein estimator (15) of $\theta:=(\theta_1,\theta_2,\ldots,\theta_m)$, for $m>2$, reduces the *overall* MSE_1 in Eq. (21). This result seems to imply that one will "do better" (in overall MSE_1 terms) by using a combined nonlinear (shrinkage) estimator, instead of estimating these means separately. What is surprising about this result is that there is no statistical reason (due to independence) to connect the inferences pertaining to the different individual means, and yet the obvious estimator (LS) is inadmissible.

As argued next, this result calls into question the appropriateness of the notion of admissibility with respect to a particular loss function, and not the judiciousness of frequentist estimation.

5. Frequentist inference and learning from data

The objectives and underlying reasoning of frequentist inference are inadequately discussed in the statistics literature. As a result, some of its key differences with Bayesian inference remain beclouded.

5.1. Frequentist approach: primary objective and reasoning

All forms of parametric frequentist inference begin with a prespecified statistical model $\mathcal{M}_\theta(\mathbf{x})=\{f(\mathbf{x};\theta),\ \theta\in\Theta\}$, $\mathbf{x}\in\mathbb{R}^n_X$. This model is chosen from the set of all possible models that could have given rise to data $\mathbf{x}_0 :=(x_1,\ldots,x_n)$, by selecting the probabilistic structure for the underlying stochastic process $\{X_t,\ t\in\mathbb{N}:=(1,2,\ldots,n,\ldots)\}$ in such a way so as to render the observed data $\mathbf{x}_0$ a "typical" realization thereof. In light of the fact that each value of $\boldsymbol{\theta}\in\Theta$ represents a different element of the family of models represented by $\mathcal{M}_\theta(\mathbf{x})$, the primary objective of frequentist inference is to learn from data about the "true" model:

$$\mathcal{M}^*(\mathbf{x})=\{f(\mathbf{x};\theta^*)\},\ \mathbf{x}\in\mathbb{R}^n_X, \tag{25}$$

where $\boldsymbol{\theta}^*$ denotes the true value of $\boldsymbol{\theta}$ in Θ. The "typicality" is testable vis-a-vis the data $\mathbf{x}_0$ using misspecification testing; see Ref. [38].

The frequentist approach relies on two modes of reasoning for inference purposes:

$$\begin{array}{ll} \textit{Factual}\ (\text{estimation, prediction}): & f(\mathbf{x};\theta^*),\ \forall\mathbf{x}\in\mathbb{R}^n_X, \\ \textit{Hypothetical}\ (\text{hypothesistesting}): & f(\mathbf{x};\theta_0),\mathrm{f}(\mathbf{x};\theta_1),\ \forall\mathbf{x}\in\mathbb{R}^n_X, \end{array} \tag{26}$$

where θ^* denotes the true value of θ in Θ, and θ_i, $i=0,1$ denote hypothesized values of θ associated with the hypotheses, H_0: $\theta_0\in\Theta_0$, H_1: $\theta_1\in\Theta_1$, where Θ_0 and Θ_1 constitute a partition of Θ.

A frequentist estimator $\hat{\theta}$ aims to pinpoint θ^*, and its optimality is evaluated by how effectively it achieves that. Similarly, a test statistic usually compares a good estimator $\hat{\theta}$ of θ with a prespecified value θ_0, but behind $\hat{\theta}$ is the value θ^* assumed to have generated data $\mathbf{x}_0$. Hence, the hypothetical reasoning is used in testing to learn about θ^*, and has nothing to do with all possible values of θ in Θ.

This contradicts misleading claims by Bayesian textbooks ([3], p. 61):

"The frequentist paradigm relies on this criterion [risk function] to compare estimators and, if possible, to select the best estimator, the reasoning being that estimators are evaluated on their long-run performance for all possible values of the parameter θ."

Contrary to this claim, the only relevant value of θ in evaluating the "optimality" of $\hat{\theta}$ is θ^*. Such misleading claims stem from an apparent confusion between the existential and universal quantifiers in framing certain inferential assertions.

The existence of θ^* can be formally defined using the *existential* quantifier:

$$\exists\theta^*\in\Theta:\quad \text{there exists a } \theta^*\in\Theta \text{ such that.} \tag{27}$$

This introduces a potential conflict between the *existential* and the *universal* quantifier "$\forall\theta\in\Theta$" because neither the decision theoretic nor the Bayesian approach explicitly invoke θ^*. Decision-theoretic and Bayesian rules are considered optimal when they minimize the expected loss $\forall\theta\in\Theta$, no matter what θ^* happens to be.

Any attempt to explain away the crucial differences between the two quantifiers can be easily scotched using elementary logic. The two quantifiers could not be more different since, using the logical connective for negation (¬), the equivalence between the two involves double negations:

$$(i)\ \exists\,\theta^*\in\Theta\Leftrightarrow\neg\forall\,\boldsymbol{\theta}\notin\Theta,\quad (ii)\forall\,\boldsymbol{\theta}\in\Theta\Leftrightarrow\neg\exists\,\theta^*\notin\Theta. \tag{28}$$

Similarly, invoking intuition to justify the quantifier $\forall\theta\in\Theta$ as innocuous and natural on the grounds that one should care about the behavior of an estimator $\hat{\theta}$ for all possible values of θ, is highly misleading. The behavior of $\hat{\theta}$, for all $\theta\in\Theta$, although relevant, is not what determines how effective a frequentist estimator is at pinpointing θ^*; what matters is its sampling behavior around θ^*. Assessing its effectiveness calls for evaluating (deductively) the sampling distribution of $\hat{\theta}$ under factual $\theta=\theta^*$, or hypothetical values θ_0 and θ_1, and not for all possible values of θ in Θ. Let's unpack the details of this claim.

5.2. Frequentist estimation

The underlying reasoning for frequentist estimation is *factual*, in the sense the optimality of an estimator is appraised in terms of its generic capacity of $\hat{\theta}_n(\mathbf{X})$ to *zero-in* on (pinpoint) the true value θ^*, whatever the sample realization $\mathbf{X}=\mathbf{x}_0$. Optimal properties like consistency, unbiasedness, full efficiency, sufficiency, etc., calibrate this generic capacity using its sampling distribution of $\hat{\theta}_n(\mathbf{X})$ evaluated under $\theta=\theta^*$ i.e., in terms of $f(\hat{\theta}_n(\mathbf{x});\theta^*)$, for $\mathbf{x}\in\mathbb{R}^n_X$. For instance, *strong consistency* asserts that as $n\to\infty$, $\hat{\theta}_n(\mathbf{X})$ will zero-in on θ^* almost surely:

$$\mathbb{P}(\lim_{n\to\infty}\hat{\theta}_n(\mathbf{X})=\theta^*)=1. \tag{29}$$

Similarly, *unbiasedness* asserts that the mean of $\hat{\theta}_n(\mathbf{X})$ is the true value θ^* :

$$E(\hat{\theta}_n(\mathbf{X}))=\theta^*. \tag{30}$$

In this sense, both of these optimal properties are defined at the point $\theta=\theta^*$. This is achieved by using *factual reasoning*, i.e., evaluating the sampling distribution of $\hat{\theta}_n(\mathbf{X})$ under the true state of Nature ($\theta=\theta^*$), without having to know θ^*. This is in contrast to using loss functions, such as Eq. (2), which are defined in terms of θ^* but are rendered nonoperational without knowing θ^*.

Example. In the case of the simple Normal model in Eq. (18), the point estimator, $\overline{X}_n$, is consistent, unbiased, fully efficient, sufficient, with a sampling distribution:

$$\overline{X}_n\sim\mathrm{N}\left(\theta,\frac{1}{n}\right). \tag{31}$$

What is not usually appreciated sufficiently is that the evaluation of that distribution is *factual*, i.e., $\theta=\theta^*$, and should formally be denoted by:

$$\overline{X}_n \overset{\theta=\theta^*}{\sim} N\left(\theta^*, \frac{1}{n}\right). \tag{32}$$

When $\overline{X}_n$ is standardized, it yields the pivotal function:

$$d(\mathbf{X};\theta) := \sqrt{n}(\overline{X}_n - \theta^*) \overset{\theta=\theta^*}{\sim} N(0,1), \tag{33}$$

whose distribution only holds for the true θ^*, and no other value. This provides the basis for constructing a $(1-\alpha)$ confidence interval (CI):

$$\mathbb{P}\left(\overline{X}_m - c_{\frac{\alpha}{2}}(\frac{1}{\sqrt{n}}) \leq \theta \leq \overline{X}_n + c_{\frac{\alpha}{2}}(\frac{1}{\sqrt{n}}); \theta=\theta^*\right) = 1-\alpha, \tag{34}$$

which asserts that the random interval $\left[\overline{X}_n - c_{\frac{\alpha}{2}}\left(\frac{s}{\sqrt{n}}\right), \overline{X}_n + c_{\frac{\alpha}{2}}\left(\frac{s}{\sqrt{n}}\right)\right]$ will cover (overlay) the true mean θ^*, *whatever that happens to be,* with probability $(1-\alpha)$, or equivalently, the error of coverage is α. Hence, frequentist evaluation of the coverage error probability depends only on the sampling distribution of $\overline{X}_n$ and is attached to random interval for all values $\theta \neq \theta^*$ without requiring one to know θ^*.

The evaluation at $\theta = \theta^*$ calls into question the decision-theoretic definition of unbiasedness:

$$E_1(\hat{\theta}_n(\mathbf{X})) = \theta, \forall \theta \in \Theta, \tag{35}$$

in the context of frequentist estimation since this assertion makes sense only when defined at $\theta=\theta^*$. Similarly, the appropriate frequentist definition of the MSE for an estimator, initially proposed by Fisher [39], is defined at the point $\theta=\theta^*$:

$$MSE(\hat{\theta}_n(\mathbf{X}); \theta^*) = E(\hat{\theta}_n(\mathbf{X}) - \theta^*)^2, \text{for } \theta^* \text{ in } \Theta. \tag{36}$$

Indeed, the well-known decomposition:

$$MSE(\hat{\theta}(\mathbf{X}); \theta^*) = Var(\hat{\theta}(\mathbf{X})) + [E(\hat{\theta}_n(\mathbf{X})) - \theta^*]^2, \text{for } \theta^* \text{ in } \Theta, \tag{37}$$

is meaningful only when defined at the point $\theta=\theta^*$ (true mean) since by definition:

$$\begin{aligned} &Var(\hat{\theta}(\mathbf{X})) = E[\hat{\theta}_n(\mathbf{X}) - \theta_m]^2,\ \theta_m = E(\hat{\theta}_n(\mathbf{X})) \\ &Bias(\hat{\theta}_n(\mathbf{X}); \theta^*) = E(\hat{\theta}_n(\mathbf{X})) - \theta^*, \end{aligned} \tag{38}$$

and thus, the variance and the bias involve only two values of θ in Θ, θ_m and θ^*, and when $\theta_m = \theta^*$ the estimator is unbiased. This implies that the apparent affinity between the MSE_1 defined in Eq. (13) and the variance of an estimator is more apparent than real because the latter makes frequentist sense only when $\theta_m = E(\hat{\theta}_n(\mathbf{X}))$ is a single point.

5.3. James-Stein estimator from a frequentist perspective

For a proper frequentist evaluation of the above James-Stein result, it is important to bring out the conflict between the *overall MSE* (14) and the factual reasoning underlying frequentist estimation. From the latter perspective, the James-Stein estimator raises several issues of concern.

First, both the least-squares $\hat{\boldsymbol{\theta}}_{LS}(\mathbf{X})$ and the James-Stein $\hat{\boldsymbol{\theta}}_{JS}(\mathbf{X})$ estimators are *inconsistent* estimators of θ since the underlying model suffers from the incidental parameter problem: there is essentially one observation (X_k) for each unknown parameter (θ_k), and as $m \to \infty$ the number of unknown parameters increases at the same rate. To bring out the futility of comparing these two estimators more clearly, consider the following simpler example.

Example. Let $\mathbf{X} := (X_1, X_2, \ldots, X_n)$ be a sample from the simple Normal model in Eq. (18). Comparing the two estimators $\hat{\boldsymbol{\theta}}_1 = X_n$ and $\hat{\boldsymbol{\theta}}_2 = \frac{1}{2}(X_1 + X_n)$ and inferring that $\hat{\boldsymbol{\theta}}_2$ is relatively more efficient than $\hat{\boldsymbol{\theta}}_1$ relative to a square loss function, i.e.,

$$\mathrm{MSE}(\hat{\theta}_2(\mathbf{X}); \theta) = 1 < \mathrm{MSE}(\hat{\theta}_1(\mathbf{X}); \theta) = \frac{1}{2}, \ \forall \theta \in \mathbb{R}, \tag{39}$$

is totally uninteresting because both estimators are inconsistent!

Second, to be able to discuss the role of admissibility in the Stein [37] result, we need to consider a consistent James-Stein estimator, by extending the original data to a panel (longitudinal) data where the sample is:

$\mathbf{X}_t := (X_{1t}, X_{2t}, \ldots, X_{mt})$, $t = 1, 2, \ldots, n$. In this case, the consistent least-squares and James-Stein estimators are:

$$\begin{aligned} &\hat{\theta}_{LS}(\mathbf{X}) = (\overline{X}_1, \overline{X}_2, \ldots, \overline{X}_m), \text{where } \overline{X}_k = \frac{1}{n}\sum_{t=1}^{n} X_{kt}, \ k = 1, 2, \ldots, m, \\ &\hat{\theta}^{+}_{JS}(\mathbf{X}) = \left(1 - \tfrac{(m-2)\sigma^2}{\|\overline{\mathbf{X}}\|^2}\right)^{+} \overline{\mathbf{X}}, \text{ where } \overline{\mathbf{X}} := (\overline{X}_1, \overline{X}_2, \ldots, \overline{X}_m). \end{aligned} \tag{40}$$

This enables us to evaluate the notion of "relatively better" more objectively.

Admissibility relative to the overall loss function in Eq. (21) introduces a trade-off between the accuracy of the estimators for individual parameters $\boldsymbol{\theta} := (\theta_1, \theta_2, \ldots, \theta_m)$ and the "overall" expected loss. The question is: "In what sense the overall MSE among a group of mean estimates provides a better measure of "error" in learning about the true values $\boldsymbol{\theta}^* := (\theta_1^*, \theta_2^*, \ldots, \theta_m^*)$?" The short answer is: it does not. Indeed, the overall MSE will be irrelevant when the primary objective of estimation is to learn from data about $\boldsymbol{\theta}^*$. This is because the particular loss function penalizes the estimator's capacity to pin-point $\boldsymbol{\theta}^*$ by trading an increase in *bias* for a decrease in the overall MSE in Eq. (21), when the latter is misleadingly evaluated over all $\boldsymbol{\theta}$ in $\Theta := \mathbb{R}^m$. That is, the James-Stein estimator flouts the primary objective of pinpointing $\boldsymbol{\theta}^*$ in favor of reducing the overall MSE $\forall \boldsymbol{\theta} \in \Theta$.

In summary, the above discussion suggests that there is nothing paradoxical about Stein's [37] original result. What is problematic is not the least-squares estimator, but the choice of "better" in terms of admissibility relative to an overall MSE in evaluating the accuracy of the estimators of $\boldsymbol{\theta}$.

5.4. Frequentist hypothesis testing

Another frequentist inference procedure one can employ to learn from data about θ^* is hypothesis testing, where the question posed is whether θ^* is close enough to some prespecified value θ_0. In contrast to estimation, the reasoning underlying frequentist testing is *hypothetical* in nature.

5.4.1. Legitimate frequentist error probabilities

For testing the hypotheses:

$H_0{:}\theta \leq \theta_0 \text{vs.} H_1{:}\theta > \theta_0$, where θ_0 is a prespecified value,

one utilizes the same sampling distribution $\overline{X}_n \sim \mathrm{N}(\theta, \frac{1}{n})$, but transforms the pivot $d(\mathbf{X};\theta) := \sqrt{n}(\overline{X}_n - \theta^*)$ into the test statistic by replacing θ^* with the prespecified value θ_0, yielding $d(\mathbf{X}) := \sqrt{n}(\overline{X}_n - \theta_0)$. However, instead of evaluating it under the factual $\theta = \theta^*$, it is now evaluated under various *hypothetical scenarios* associated with H_0 and H_1 to yield two types of (hypothetical) sampling distributions:

(I) $d(\mathbf{X}) := \sqrt{n}(\overline{X}_n - \theta_0) \overset{\theta=\theta^*}{\sim} \mathrm{N}(0, 1),$

(II) $d(\mathbf{X}) := \sqrt{n}(\overline{X}_n - \theta_0) \overset{\theta=\theta^*}{\sim} \mathrm{N}(\delta_1, 1),\ \delta_1 = \sqrt{n}(\theta_1 - \theta_0) \text{for } \theta_1 > \theta_0.$

In both cases, (I) and (II), the underlying reasoning is hypothetical in the sense that the factual in Eq. (33) is replaced by hypothesized values of θ, and the test statistic $d(\mathbf{X})$ provides a standardized distance between the hypothesized values (θ_0 or θ_1) and θ^* the true θ, assumed to underlie the generation of the data $\mathbf{x}_0$, yielding $d(\mathbf{x}_0)$. Using the sampling distribution in (I), one can define the following legitimate error probabilities:

$$\begin{aligned} \textbf{significance level:} &\quad \mathbb{P}(d(\mathbf{X}) > c_\alpha; H_0) = \alpha, \\ \textbf{p-value:} &\quad \mathbb{P}(d(\mathbf{X}) > d(\mathbf{x}_0); H_0) = p(\mathbf{x}_0). \end{aligned} \tag{41}$$

Using the sampling distribution in (II), one can define:

$$\begin{aligned} \textbf{type II error prob.:} &\quad \mathbb{P}(d(\mathbf{X}) \leq c_\alpha; \theta = \theta_1) = \beta(\theta_1), \text{ for } \theta_1 > \theta_0, \\ \textbf{power:} &\quad \mathbb{P}(d(\mathbf{X}) > c_\alpha; \theta = \theta_1) = \rho(\theta_1), \text{ for } \theta_1 > \theta_0. \end{aligned} \tag{42}$$

It can be shown that the test T_α, defined by the test statistic $d(\mathbf{X})$ and the rejection region $C_1(\alpha) = \{\mathbf{x} : d(\mathbf{x}) > c_\alpha\}$, constitutes a uniformly most powerful (UMP) test for significance level α; see Ref. [9]. The type I [II] error probability is associated with test T_α erroneously rejecting [accepting] H_0. The type I and II error probabilities evaluate the generic capacity [whatever the sample realization $\mathbf{x} \in \mathbb{R}^n$] of a test to reach correct inferences. Contrary to Bayesian claims, these error probabilities have nothing to do with the temporal or the physical dimension of the long-run metaphor associated with repeated samples. The relevant feature of the long-run metaphor is the repeatability (in principle) of the DGM represented by $\mathcal{M}_\theta(\mathbf{x})$; this feature can be easily operationalized using computer simulation; see Ref. [40].

The key difference between the significance level α and the p-value is that the former is a *pre-data* and the latter a *post-data* error probability. Indeed, the p-value can be viewed as the smallest significance level α at which H_0 would have been rejected with data $\mathbf{x}_0$. The legitimacy of postdata error probabilities underlying the hypothetical reasoning can be used to go beyond the N-P accept/reject rules and provide an evidential interpretation pertaining to the discrepancy γ from the null warranted by data $\mathbf{x}_0$; see Ref. [41].

Despite the fact that frequentist testing uses hypothetical reasoning, its main objective is also to learn from data about the true model $\mathcal{M}^*(\mathbf{x})=\{f(\mathbf{x};\theta^*)\}$, $\mathbf{x}\in\mathbb{R}^n_X$. This is because a test statistic like $d(\mathbf{X}):=\sqrt{n}(\overline{X}_n-\theta_0)$ constitutes nothing more than a scaled distance between θ^* [the value behind the generation of $\overline{x}_n$], and a hypothesized value θ_0, with θ^* being replaced by its "best" estimator $\overline{X}_n$.

6. Revisiting loss and risk functions

The above discussion raises serious doubts about the role of loss functions and admissibility in evaluating learning from data $\mathbf{x}_0$ about $\boldsymbol{\theta}^*$. To understand why the decision-theoretic framing misrepresents the frequentist approach, one needs to consider the role of loss functions in statistical inference more generally.

6.1. Where do loss functions come from?

A closer scrutiny of the decision-theoretic set up reveals that the loss function needs to invoke "information from sources other than the data," which is usually not readily available. Indeed, such information is available in very restrictive situations, such as acceptance sampling in quality control. In light of that, a proper understanding of the intended scope of statistical inference calls for distinguishing the special cases where the loss function is part and parcel of the available substantive information from those that no such information is either relevant or available.

Tiao and Box [25], p. 624, reiterated Fisher's [42] distinction:

"Now it is undoubtedly true that on the one hand that situations exist where the loss function is at least approximately known (for example, certain problems in business) and sampling inspection are of this sort. ... On the other hand, a vast number of inferential problems occur, particularly in the analysis of scientific data, where there is no way of knowing in advance to what use the results of research will subsequently be put."

Cox [43] went further and questioned this framing even in cases where the inference might involve a decision:

"The reasons that the detailed techniques [decision-theoretic] seem of fairly limited applicability, even when a fairly clear cut decision element is involved, may be (i) that, except in such fields as control theory and acceptance sampling, a major contribution of statistical technique is in presenting the evidence in incisive form for discussion, rather than in providing mechanical presentation for the final decision. This is especially the case when a single major decision is involved. (ii) The central difficulty may be in formulating the elements required for the quantitative analysis, rather than in combining these elements via a decision rule." (p. 45)

Another important aspect of using loss functions in inference is that in practice they seem to be an add-on to the inference itself since they bring to the problem the information other than the data. In particular, the same statistical inference problem can give rise to very different decisions/actions depending on one's loss function. To illustrate that consider an example from [44]:

"… consider the case of a new drug whose effects are studied by a research scientist attached to the laboratory of a pharmaceutical company. The conclusion of the study may have different bearings on the action to be taken by (a) the scientist whose line of further investigation would depend on it, (b) the company whose business decisions would determined by it, and (c) the Government whose policies as to health care, drug control, etc., would take shape on that basis." (p. 72)

In practice, each one of these different agents is likely to have a very different loss function, but their inferences should have a common denominator: the scientific evidence pertaining to θ^*, the true θ, that stems solely from the observed data.

6.2. Decisions vs. inferences

The above discussion brings out the crucial distinction between a "decision" and an "inference" stemming from data $\mathbf{x}_0$. Even before Wald [5] introduced the decision-theoretic perspective, Fisher [42] perceptively argued:

"In the field of pure research no assessment of the cost of wrong conclusions, or of delay in arriving at more correct conclusions can conceivably be more than a pretence, and in any case such an assessment would be inadmissible and irrelevant in judging the state of the scientific evidence." (pp. 25–26)

Tukey (1960) echoed Fisher's view by contrasting decisions vs. inferences:

"Like any other human endeavor, science involves many decisions, but it progresses by the building up of a fairly well established body of knowledge. This body grows by the reaching of conclusions — by acts whose essential characteristics differ widely from the making of decisions. Conclusions are established with careful regard to evidence, but without regard to consequences of specific actions in specific circumstances." (p. 425)

Hacking [45] brought out the key difference between an "inference pertaining to evidence" for or against a hypothesis, and a "decision to do something" as a result of an inference:

"… to conclude that an hypothesis is best supported is, apparently, to decide that the hypothesis in question is best supported. Hence it is a decision like any other. But this inference is fallacious. Deciding that something is the case differs from deciding to do something. ... Hence deciding to do something falls squarely in the province of decision theory, but deciding that something is the case does not." (p. 31)

This issue was elaborated upon by Birnbaum [15], p. 19:

"Two contrasting interpretations of the decision concept are formulated: *behavioral*, applicable to "decisions" in a concrete literal sense as in acceptance sampling; and *evidential*, applicable to "decisions" such as "reject H_0" in a research context, where the pattern and strength of statistical evidence concerning statistical hypotheses is of central interest."

6.3. Loss functions vs. inherent distance functions

The notion of a loss function stemming from "information other than the data" raises another source of potential conflict. This stems from the fact that within each statistical model $\mathcal{M}_\theta(\mathbf{x})$ there exists an *inherent* statistic distance function, often relating to the log-likelihood and the score function, which constitutes information contained in the data; see Ref. [46].

It is well known that when the distribution underlying $\mathcal{M}_\theta(\mathbf{x})$ is normal, the *inherent distance function* for comparing estimators of the mean (θ) is the square:

$$ND(\hat{\theta}_n(\mathbf{X}); \theta^*) = (\hat{\theta}_n(\mathbf{X}) - \theta^*)^2. \tag{43}$$

On the other hand, when the distribution is *Laplace*, the relevant statistical distance function is the *absolute distance* (see Ref. [47]):

$$AD(\hat{\theta}_n(\mathbf{X}); \theta^*) = |\hat{\theta}_n(\mathbf{X}) - \theta^*|. \tag{44}$$

Similarly, when the distribution underlying $\mathcal{M}_\theta(\mathbf{x})$ is *uniform*, the inherent distance function is:

$$SUP(\hat{\theta}_n(\mathbf{X}); \theta^*) = \sup_{\mathbf{x}\in\mathbb{R}^n_X} |\hat{\theta}_n(\mathbf{x}) - \theta^*|. \tag{45}$$

Note that these distance functions are defined at the point $\theta{=}\theta^*$ and not for all θ in Θ, as traditional loss functions.

The dilemma facing a Bayesian or a decision-theoretic statistician is to decide when it makes sense to override the MLE and select the optimal rule stemming from an externally given loss function. The dilemma is not as trivial as it might seem at first sight for two reasons. First, the key difference between the two is that the assumptions of the likelihood function $L(\boldsymbol{\theta})$ are testable vis-a-vis the data, but those underlying the loss function are not. Second, the likelihood function renders the notion of efficiency "global," *full efficiency*, in terms of Fisher's information:

$$CR(\boldsymbol{\theta}^*){=}I_n^{-1}(\boldsymbol{\theta}^*),I_n(\boldsymbol{\theta}^*) := E\left(-\frac{\partial^2 \ln L(\boldsymbol{\theta})}{\partial\boldsymbol{\theta}\partial\boldsymbol{\theta}^T}\right). \tag{46}$$

Hence, the optimality of an estimator can be affirmed using testable information comprising the statistical model $\mathcal{M}_\theta(\mathbf{x})$. This is in direct contrast with admissibility, which is a property defined in terms of "local" efficiency—*relative* to a loss function—based on external (nontestable) information.

6.4. Acceptance sampling vs. learning from data

Let us bring out the key features of a situation where the above decision-theoretic set up makes perfectly good sense. This is the situation Fisher [12] called *acceptance sampling,* such as an industrial production process where the objective is quality control, i.e., to make a decision pertaining to shipping sub-standard products (e.g., nuts and bolts) to a buyer using the expected loss/gain as the ultimate criterion.

In an acceptance sampling context, the $\mathrm{MSE}(\hat{\theta}(\mathbf{X});\theta)$, or some other risk function, are relevant because they evaluate genuine losses associated with a decision related to the choice of an estimate $\hat{\theta}(\mathbf{x}_0)$, say the cost of the observed percentage of defective products, but that has nothing to do with type I and II error probabilities.

Acceptance sampling differs from a scientific enquiry in two crucial respects:

a. The primary aim is to use statistical rules to minimize the expected loss associated with "a decision."

b. The sagacity of all actions is determined by the respective "losses" stemming from "*relevant information other than the data* ([32], p. 251)."

c. The trade-off between the two types of error probabilities is determined by the risk function itself and not by any endeavor to learn from data about θ^*. Indeed, the learning is deliberately undermined by certain loss function such as the *overall* MSE (14) that favor biased estimators of the James-Stein type.

The key difference between acceptance sampling and a scientific inquiry is that the primary objective of the latter is *not* to minimize expected loss (costs and utility) associated with different values of $\boldsymbol{\theta}\in\Theta$, but to use data $\mathbf{x}_0$ to learn about the "true" model (17). The two situations are drastically different mainly because the key notion of a "true $\boldsymbol{\theta}$" calls into question the above acceptance sampling set up. Indeed, the loss function being defined "$\forall\boldsymbol{\theta}\in\Theta$," will penalize $\boldsymbol{\theta}^*$, since there is no reason to expect that the highest ranked $\boldsymbol{\theta}$ would coincide with $\boldsymbol{\theta}^*$, unless by accident.

The extreme relativism of loss function optimality renders decision-theoretic and Bayes rules highly vulnerable to abuse. In practice, one can justify any estimator as optimal, however lame in terms of other criteria, by selecting an "appropriate" loss function.

Example 1. Consider a manufacturer of high precision bolts and nuts who has information that the buyer only checks the first and last box for quality control when accepting an order. This suggests that to minimize losses, stemming from the return of its products as defective, an appropriate loss function might be:

$$L(\mathbf{X};\theta)=\left([(X_1+X_n)/2]-\theta\right)^2,\ \theta\in(0,1). \tag{47}$$

From the *acceptance sampling* perspective, the "optimal" estimator $\tilde{\theta} = (X_1 + X_n)/2$ is excellent because it minimizes the expected losses, but it is a terrible estimator for pinpointing θ^* because it is inconsistent!

Consider a more general case where acceptance sampling resembles hypothesis testing in so far as final products are randomly selected for inspection during the production process. In such a situation the main objective can be viewed as operationalizing the probabilities of false acceptance/rejection with a view to minimize the expected losses. The conventional wisdom has been that this situation is similar enough to Neyman-Pearson (N-P) testing to render the latter as the appropriate framing for the decision to ship this particular batch or not. However, a closer look at some of the examples used to illustrate such a situation [48], reveals that the decisions are driven exclusively by the risk function and not by any quest to learn from data about the true $\boldsymbol{\theta}^*$. For instance, N-P way of addressing the trade-off between the two types of error probabilities, fixing α to a small value and seek a test that minimizes the type II error probability, seems utterly irrelevant in such a context. One can easily think of a loss function where the "optimal" trade-off calls for a much larger type I than type II error probability. As argued in Ref. [14]:

"Wald's decision theory … has given up fixed probability of errors of the first kind, and has focused on gains, losses or regrets." (p. 433)

Indeed, Wald [5] was the first to highlight that the decision-theoretic notion of "optimality" revolves around a particular loss function:

"The "best" system of regions of acceptance … will depend only on the weight function of the errors." ([5], p. 302)

Given the crucial differences in [a]–[c], one can make a strong case that the objectives and the underlying reasoning of acceptance sampling are drastically different from those pertaining to learning from data in a scientific context.

6.5. Is expected loss a legitimate frequentist error?

The key question is: what do expected losses and traditional frequentist errors, such as bias, MSE and the type I–II errors, have in common, if anything?

First, they stem directly from the statistical model $\mathcal{M}_\theta(\mathbf{x})$ since the underlying sampling distributions of estimators, test statistics, and predictors are derived *exclusively from the distribution of the sample* $f(\mathbf{x};\boldsymbol{\theta})$ through Eq. (7). In this sense, the relevant error probabilities are directly related to statistical information pertaining to the data as summarized by the statistical model $\mathcal{M}_\theta(\mathbf{x})$ itself.

Second, they are attached to a particular frequentist inference procedure as they relate to a relevant inferential claim. These error probabilities calibrate the effectiveness of inference procedures in learning from data about the true statistical model $\mathcal{M}^*(\mathbf{x})=\{f(\mathbf{x};\boldsymbol{\theta}^*)\}$, $\mathbf{x}\in\mathbb{R}^n_X$.

In light of these features, the question is: "in what sense a risk function could potentially represent relevant frequentist errors?" That argument that the risk function represents legitimate frequentist errors because it is derived by taking expectations with respect to $f(\mathbf{x};\boldsymbol{\theta})$, $\mathbf{x}\in\mathbb{R}^n_X$ [3], is misguided for two reasons.

a. The relevant errors in estimation, including the bias $E(\hat{\theta}_n(\mathbf{X})) - \theta^*$ and MSE $E(\hat{\theta}_n(\mathbf{X}) - \theta^*)^2$, are evaluated with respect to $f(\mathbf{x}; \theta^*)$, $\mathbf{x} \in \mathbb{R}^n_X$, by invoking factual reasoning; θ^* denotes the state of Nature. Wald's [5] original loss function in Eq. (2) represents an interesting case because it is defined in terms of θ^*, which renders it nonoperational when evaluated for all θ in Θ, since θ^* is unknown in practice. In contrast, the errors associated with the bias and MSE are rendered operational by the factual reasoning fashioned to forgo knowing θ^*.

b. The expected losses stemming from the risk function $R(\theta,\hat{\theta})$ are attached to particular values of θ in Θ. Such an assignment is in direct conflict with all the above legitimate error probabilities that are attached to the inference procedure itself, and never to the particular values of θ in Θ. The expected loss assigned to each value of θ in Θ has nothing to do with learning from data about θ^*. Indeed, the risk function will penalize a procedure for pinpointing θ^* since the latter is unknown in practice. This is in direct conflict with the main objective of frequentist estimation but in sync with "acceptance sampling," where the objective of the inference has everything to do with expected losses.

7. Summary and conclusions

'The paper makes a case for Fisher's [12, 42] assertions concerning the appropriateness of the decision-theoretic framing for "acceptance sampling" and its inappropriateness for frequentist inference. A closer look at this framing reveals that it is congruent with the Bayesian approach because it supplements the posterior distribution with a theory of optimal inference. Decision-theoretic and Bayesian rules are considered optimal when they minimize the expected loss for all possible values of θ $[\forall\theta \in \Theta]$, irrespective of what the true value θ^* happens to be. In contrast, the theory of optimal frequentist inference revolves around the true value θ^*, since it depends entirely on the capacity of the procedure to pinpoint θ^*. The frequentist approach relies on factual (estimation and prediction), as well as hypothetical (testing) reasoning, both of which revolve around the existential quantifier $\exists\theta^* \in \Theta$. The inappropriateness of the quantifier $\forall\theta \in \Theta$ calls into question the relevance of *admissibility* as a minimal property for frequentist estimators. A strong case can be made that the relevant minimal property for frequentist estimators is *consistency*. In addition, full efficiency provides the relevant measure of an estimator's finite sample efficiency (accuracy) in pinpointing θ^*. Both of these properties stem from the underlying statistical model $\mathcal{M}_\theta(\mathbf{x})$, in contrast to admissibility which relies on loss functions based on information other than the data.

It is argued that Stein's [36] result stems from the fact that admissibility introduces a trade-off between the accuracy of the estimator in pinpointing θ^* and the "overall" expected loss. That is, the James-Stein estimator achieves a higher overall MSE by blunting the capacity of a frequentist estimator to pinpoint θ^* Why would a frequentist care about the overall MSE defined for all θ in Θ? After all, expected losses are not legitimate errors similar to bias and MSE (when properly defined), as well as coverage, type I and II errors. The latter are attached to the frequentist procedures themselves to calibrate their capacity to achieve learning from data about θ^*. In contrast, expected losses are assigned to different values of θ in Θ, using information other than the data.

Author details

Aris Spanos

Address all correspondence to: aris@vt.edu

Department of Economics, Virginia Tech, Blacksburg, VA, USA

References

[1] Wald, A. Statistical Decision Functions. NY: Wiley; 1950.

[2] Berger, J.O. Statistical Decision Theory and Bayesian Analysis, 2nd ed., NY: Springer; 1985.

[3] Robert, C.P. The Bayesian Choice: From Decision-Theoretic Foundations to Computational Implementation, 2nd ed. NY: Springer; 2001.

[4] O'Hagan, A. Bayesian Inference, London: Edward Arnold; 1994.

[5] Wald, A. Contributions to the theory of statistical estimation and testing hypotheses. Annals of Mathematical Statistics. 1939; **10**: 299–326.

[6] Neyman, J. Outline of a theory of statistical estimation based on the classical theory of probability. Philosophical Transactions of the Royal Society of London, Series A, 1937; **236**: 333–380.

[7] Neyman, J. Lectures and Conferences on Mathematical Statistics and Probability, 2nd ed. Washington: U.S. Department of Agriculture; 1952.

[8] Neyman, J. Foundations of Behavioristic Statistics. Godambe, V. and Sprott, D. eds., Foundations of Statistical Inference. Toronto: Holt, Rinehart and Winston of Canada; 1971: pp. 1–13.

[9] Lehmann, E.L. Testing Statistical Hypotheses. NY: Wiley; 1959.

[10] LeCam, L. Asymptotic Methods in Statistical Decision Theory. NY: Springer; 1986.

[11] Neyman, J.A. Selection of Early Statistical Papers by J. Neyman. Moss Landing, CA: University of California Press; 1967.

[12] Fisher, R.A. Statistical methods and scientific induction. Journal of the Royal Statistical Society, B, 1955; **17**: 69–78.

[13] Cox, D.R. Some problems connected with statistical inference. The Annals of Mathematical Statistics. 1958; **29**: 357–372.

[14] Tukey, J.W. Conclusions vs Decisions. Technometrics. 1960; **2**: 423–433.

[15] Birnbaum, A. The Neyman-Pearson theory as decision theory, and as inference theory; with a criticism of the Lindley-Savage argument for Bayesian Theory. Synthese. 1977; **36**: 19–49.

[16] Ferguson, T.S. Mathematical Statistics: A Decision Theoretic Approach. London: Academic Press. 1967.

[17] Wasserman, L. All of Statistics. NY: Springer; 2004.

[18] Bansal, A.K. Bayesian Parametric Inference. Oxford: Alpha Science. 2007.

[19] Ferguson, T.S. Development of the Decision Model. On the History of Statistics and Probability, edited by D. B. Owen. NY: Marcel Dekker. 1976; ch. 16.

[20] Wald, A. An essentially complete family class of admissible decision functions. Annals of Mathematical Statistics. 1947; **18**: 549–555.

[21] LeCam, L. An extension of Wald's theory of statistical decision functions. Annals of Mathematical Statistics. 1955; **26**: 69–81.

[22] Spanos, A. and Mayo, D.G. Error statistical modeling and inference: where methodology meets ontology. Synthese. 2015; **192**: 3533–3555.

[23] Reid, N. Statistical Sufficiency. International Encyclopedia of the Social & Behavioral Sciences, edited by Wright J.D., 2nd edition, Vol 23. Oxford: Elsevier. 2015; pp. 418–422.

[24] Lindley, D.V. Introduction to Probability and Statistics from a Bayesian Viewpoint, Part 2: Inference. Cambridge: Cambridge University Press; 1965.

[25] Tiao, G.C. and Box, G.E.P. Some comments on "Bayes" estimators. Studies in Bayesian Econometrics and Statistics, In Honor of Leonard Savage J., edited by Fienberg S.E. and Zellner A. Amsterdam: North-Holland. 1975; pp. 619–626.

[26] Spanos, A. Revisiting Bayes' Rule and Evidence: Notional Events vs. Real Data. Virginia Tech, working paper; Blacksburg, VA. 2015.

[27] Schervish, M.J. Theory of Statistics. New York, NY: Springer-Verlag; 1995.

[28] Srivastava, M.K., Khan A.H. and Srivastava, N. Statistical Inference: Theory of Estimation, Delhi, India: PHI Learning; 2014.

[29] Rubin, D.B. Bayesianly justifiable and relevant frequency calculation for the applied statistician. Annals of Statistics. 1984; **12**: 1151–1172.

[30] Gelman, A., Carlin J.B. and Rubin, D.B., Bayesian Data Analysis, 2nd edition, London: Chapman & Hall; 2004.

[31] Savage, L.J. On rereading Fisher, R.A. The Annals of Statistics. (1976); **4**(3): 441–500.

[32] Cox, D.R. and Hinkley, D.V. Theoretical Statistics. London: Chapman & Hall; 1974.

[33] Lehmann, E.L. Specification Problems in the Neyman-Pearson-Wald Theory. Statistics: An Appraisal, edited by David, H.A. and David, H.T., IA: The Iowa State University Press; 1984; 425–436.

[34] Efron, B. and Morris, C.N. Stein's estimation rule and its competitors–an empirical Bayes approach. Journal of the American Statistical Association. 1973; **68**: 117–130.

[35] Saleh, A.K. Md. E. Theory of Preliminary Test and Stein-Type Estimation with Applications. NY: Wiley; 2006.

[36] Stein, C. Inadmissibility of the usual estimator for the mean of a multivariate distribution. Proceedings of the Third Berkeley Symposium on Mathematical Statistics and Probability. 1956; **1**: 197–206.

[37] James, W. and Stein, C. Estimation with quadratic loss. Proceedings of the Fourth Berkeley Symposium on Mathematical Statistics and Probability. 1961; **1**: 361–379.

[38] Spanos, A. Where Do Statistical Models Come From? Revisiting the Problem of Specification. The Second Erich L. Lehmann Symposium, edited by Rojo J., Lecture Notes-Monograph Series, vol. 49, Hayward CA: Institute of Mathematical Statistics. 2006; pp. 98–119.

[39] Fisher, R.A. A Mathematical examination of the methods of determining the accuracy of an observation by the mean error, and by the mean square error. Monthly Notices of the Royal Astronomical Society. 1920; **80**: 758–770.

[40] Spanos, A. A frequentist interpretation of probability for model-based inductive inference. Synthese. 2013; **190**: 1555–1585.

[41] Mayo, D.G. and Spanos, A. Severe testing as a basic concept in a Neyman-Pearson philosophy of induction. The British Journal for the Philosophy of Science; 2006; **57**: 323–357.

[42] Fisher, R.A. The Design of Experiments. Edinburgh: Oliver and Boyd; 1935.

[43] Cox, D.R. Foundations of statistical inference: the case for eclecticism. Australian Journal of Statistics. 1978; **20**: 43–59.

[44] Chatterjee, S.K. Statistical Thought: A Perspective and History. Oxford: Oxford University Press. 2002.

[45] Hacking, I. Logic of Statistical Inference. Cambridge: Cambridge University Press; 1965.

[46] Casella, G. and Berger, R.L. Statistical Inference, 2nd ed. CA: Duxbury. 2002.

[47] Shao, J. Mathematical Statistics. 2nd ed. NY: Springer. 2003.

[48] Silvey, S.D. Statistical Inference, London: Chapman & Hall. 1975.

4

Gamma-Kumaraswamy Distribution in Reliability Analysis: Properties and Applications

Indranil Ghosh and Gholamhossein G. Hamedani

Abstract

In this chapter, a new generalization of the Kumaraswamy distribution, namely the gamma-Kumaraswamy distribution is defined and studied. Several distributional properties of the distribution are discussed in this chapter, which includes limiting behavior, mode, quantiles, moments, skewness, kurtosis, Shannon's entropy, and order statistics. Under the classical method of estimation, the method of maximum likelihood estimation is proposed for the inference of this distribution. We provide the results of an analysis based on two real data sets when applied to the gamma-Kumaraswamy distribution to exhibit the utility of this model.

Keywords: gamma-Kumaraswamy distribution, Renyi's entropy, reliability parameter, stochastic ordering, characterizations

1. Introduction

The generalization of a distribution by mixing it with another distribution over the years has provided a mathematical based way to model a wide variety of random phenomena statistically. These generalized distributions are effective and flexible models to analyze and interpret random durations in a possibly heterogeneous population. In many situations, observed data may be assumed to have come from such a mixture population of two or more distributions.

Two parameter gamma and a two parameter Kumaraswamy are most popular distribution for analyzing any lifetime data. Gamma distribution is a well-known distribution, and it has several desirable properties [1].

A serious limitation of the gamma distribution, however, is that the distribution function (or survival function) is not available in a closed form if the shape parameter is not an integer, thereby it requires some numerical methods to evaluate these quantities. As a consequence, this distribution is less attractive as compared to Ref. [2], which has nice tractable distribution function, survival function and hazard function. In this paper, we consider a four parameter gamma-Kumaraswamy distribution. It is observed that it has many properties which are quite similar to those of a gamma distribution, but it has an explicit expression for the distribution function or the survival functions. The major motivation of this chapter is to introduce a new family of distributions, make a comparative study of this family with respect to a Kumaraswamy family and a gamma family and provide the practitioner with an additional option, with a hope that it may have a 'better fit' compared to a gamma family or Kumaraswamy family in certain situations. It is noteworthy to note that the gamma-Kumaraswamy distribution is a generalization of Kumaraswamy distribution with the property that it can exhibit various shapes. (**Figure 1**). This provides more flexibility to the gamma-Kumaraswamy distribution in comparison with Kumaraswamy distribution in modeling different data sets. The property of left-skewness is a rare characteristic as it is not enjoyed by several generalizations of Kumaraswamy distribution. Our proposed model is different from that of Ref. [3], where the authors have proposed a generalized gamma-generated distribution with an extra positive parameter for any continuous baseline G distribution.

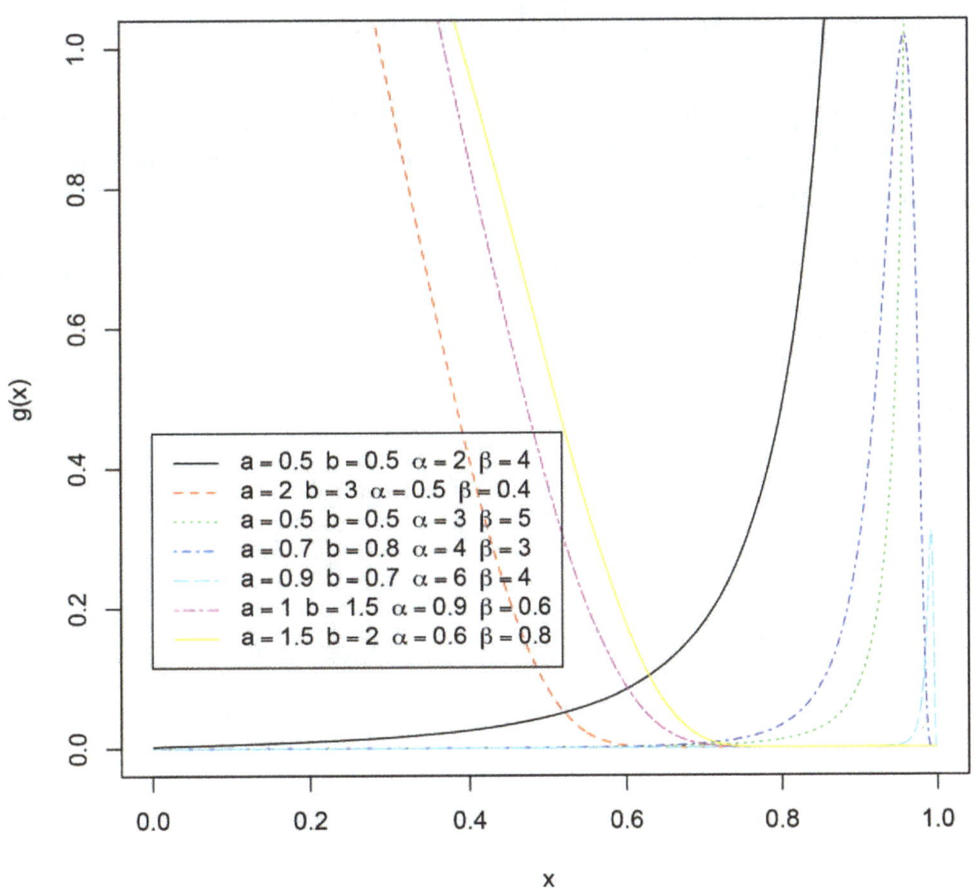

Figure 1. GK density plot for some specific parameter values.

The rest of the paper is organized as follows. In Section 2, we propose the gamma-Kumaraswamy distribution [GK(α, β, a, b)]. In Section 3, we study various properties of the GK(α, β, a, b) including the limiting behavior, transformation, and the mode. In Section 4, the moment generating function, the moments and the mean deviations from the mean and the median, and Renyi's entropy are studied. In Section 5, we consider the maximum likelihood estimation of the GK(α, β, a, b). In Section 6, we provide an expression for the reliability parameter for two independent GK(α, β, a, b) with different choices for the parameters α and β but for a fixed choice of the two shape parameters of Kumaraswamy distribution. In Section 7, discussion is made for the moment generating function of the r-th order statistic and also the limiting distribution of the sample minimum and the sample maximum for a random sample of size n drawn from GK(α, β, a, b). An application of GK(α, β, a, b) is discussed in Section 8. Certain characterizations of GK(α, β, a, b) are presented in Section 9. In Section 10, some concluding remarks are made.

2. The gamma-Kumaraswamy distribution

We consider the following class of gamma-X class of distributions, for which, the parent model being

$$f(x) = \frac{1}{\Gamma(\alpha)\beta^{\alpha}} \frac{g(x)}{\overline{G}^{2}(x)} \exp\left(-\frac{g(x)}{\beta\overline{G}(x)}\right)\left(\frac{G(x)}{\overline{G}(x)}\right)^{\alpha-1}, \quad x > 0, \tag{1}$$

where α, β are positive parameters. Also, $g(x)[G(x)]$ is the density function [cumulative distribution function] of the random variable X. Furthermore, $\overline{G}(x)$ is the survival function of the associated random variable X.

If X has density Eq. (1), then the random variable $W = \frac{G(x)}{\overline{G}(x)}$ has a gamma distribution with parameters α, β. The reverse happens to be true as well. Here, we consider G(.) to be the cdf of a Kumaraswamy distribution with parameters a, b. Then, the cdf of the gamma-Kumaraswamy (hereafter GK) reduces to

$$F(x) = \int_0^{\frac{1-(1-x^a)^b}{(1-x^a)^b}} \frac{e^{-w/\beta} w^{\alpha-1}}{\Gamma(\alpha)\beta^{\alpha}} dw = \gamma_1\left(\alpha, \frac{1-(1-x^a)^b}{\beta(1-x^a)^b}\right), \quad 0 < x < 1. \tag{2}$$

where $\gamma_1(\alpha, z) = \frac{\Gamma(\alpha,\, z)}{\Gamma(\alpha)}$ with $\Gamma(\alpha, x) = \int_0^x u^{\alpha-1} e^{-u} du$ is the regularized incomplete gamma function. So the density and hazard functions corresponding to Eq. (2) are given, respectively, by

$$f(x) = \frac{ab \exp\left(-\frac{1-(1-x^a)^b}{\beta(1-x^a)^b}\right)}{\Gamma(\alpha)\beta^{\alpha}} \frac{1}{(1-x^a)^2} \left(\frac{1-(1-x^a)^b}{\beta(1-x^a)^b}\right)^{\alpha-1} x^{a-1}, \quad 0 < x < 1, \tag{3}$$

and

$$h_F(x) = \frac{\left((1-x^a)^{-b}-1\right)^{\alpha-1} abx^{a-1}(1-x^a)^{b-1}\exp\left(-\beta^{-1}((1-x^a)^{-b}-1)\right)}{\beta^{\alpha}(1-x^a)^{b+1}\left(1-\Gamma\left(\alpha,\frac{1-(1-x^a)^b}{\beta(1-x^a)^b}\right)\right)}. \quad (4)$$

The percentile functions for GK distribution: The p th percentile x_p is defined by $F(x_p) = p$. From Eq. (2), we have $\gamma_1\left(\alpha,\frac{1-(1-x^a)^b}{\beta(1-x^a)^b}\right) = p$. Define $Z_p = \frac{1-(1-x^a)^b}{\beta(1-x^a)^b}$, then $Z_p = \gamma_1^{-1}(\alpha,p)$, where γ_1^{-1} is the inverse of regularized incomplete gamma function. Hence, $x_p = \left(1-\left(\beta(1+Z_{1-p})\right)^{1/b}\right)^{1/a}$.

In the density equation (3), a, b, and α are shape parameters and β is the scale parameter. It can be immediately verified that Eq. (3) is a density function. Plots of the GK density and survival rate function for selected parameter values are given in **Figures 1** and **2**, respectively.

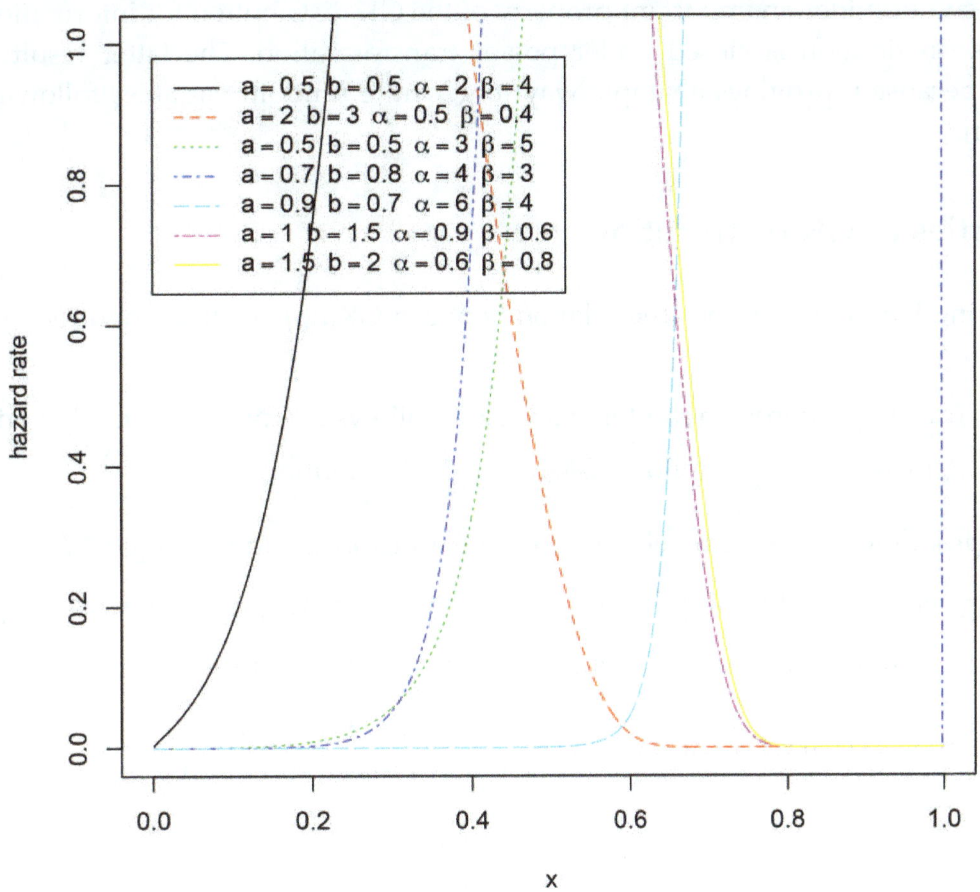

Figure 2. GK hazard rate function plot for some specific parameter values.

If X~$GK(a, b, \alpha, \beta)$, then the survival function of X, $S(x)$ will be

$$1-\gamma_1\left(\alpha,\frac{1-(1-x^a)^b}{\beta(1-x^a)^b}\right). \quad (5)$$

We simulate the GK distribution by solving the nonlinear equation

$$(1{-}u){-}\gamma_1\left(\alpha,\frac{1{-}(1{-}x^a)^b}{\beta(1{-}x^a)^b}\right)=0, \tag{6}$$

where u has the uniform (0,1) distribution. Some facts regarding the GK distribution are as follows:

- If X~$GK(a, b, \alpha, \beta)$, then X^m~$GK(a, b, \alpha, \beta)$, $\forall m\neq 0$.
- Also, we have the following important result: If X~$GK(1, b, \alpha, \beta)$, then $X^{1/a}$~$GK(a, b, \alpha, \beta)$, $\forall a\neq 0$.
- The GK distribution does not possess the reproductive property. In other words, if for any two X_1~$GK(a_1,b_1,\alpha_1,\beta_1)$ and X_2~$GK(a_2,b_2,\alpha_2,\beta_2)$, then the distribution of the sum $S = X_1 + X_2$ will not be a GK.

The first result provides an important property of the GK distribution for information analysis is that this distribution is closed under power transformation. The latter result is equally important because it provides a simple way to generate random variables following the GK distribution.

3. Properties of GK distribution

The following lemma establishes the relation between GK(α, β, a, b) distribution and gamma distribution.

Lemma 1. (Transformation): If a random variable X follows a gamma distribution with parameters α and β, then $Y=\frac{1{-}(1{-}X^a)^b}{(1{-}X^a)^b}$ follows GK(α, β, a, b) distribution.

Proof. The proof follows immediately by using the transformation technique. W

The limiting behaviors of the GK pdf and its hazard function are given in the following theorem.

Theorem 1. The limits GK density function, $f(x)$, and the hazard function, $h_F(x)$, are given by

$$\lim_{x\to 0^+} f(x)=\lim_{x\to 0^+} h_f(x)=\begin{cases}0, & a>1, b>1, \alpha>1\\ \infty, & \min\{a,b\}<1, \alpha<1,\end{cases} \tag{7}$$

$$\lim_{x\to\infty} f(x)=\lim_{x\to\infty} h_f(x)=\begin{cases}0, & b>0, \alpha<1\\ \infty, & b<0, \alpha>1.\end{cases} \tag{8}$$

Proof. Straightforward and hence omitted. W

Theorem 2. The mode of the GK distribution is the solution of the equation $k(x)=0$, where

$$k(x)=(a{-}1){-}\frac{2x^a}{(1{-}x^a)}+\frac{abx^a}{(1{-}x^a)^b}\left(\beta^{-1}+\left(\frac{1{-}(1{-}x^a)^b}{\beta(1{-}x^a)^b}\right)^{-1}\right). \tag{9}$$

Proof. The derivative of $f(x)$ in Eq. (3) can be written as

$$\frac{\partial}{\partial x}f(x)=\frac{1}{\beta^{\alpha}\Gamma(\alpha)}\frac{abx^{a-2}}{(1-x^{a})^{2}}\exp\left(-\beta^{-1}((1-x^{a})^{-b}-1)\right)\left(\frac{1-(1-x^{a})^{b}}{\beta(1-x^{a})^{b}}\right)^{\alpha-1}k(x). \tag{10}$$

The critical values of Eq. (10) are the solutions of $k(x)=0$. W

Next, we discuss the IFR and/or DFR property of the hazard function for the GK distribution. For this, we will consider the result of Lemma 1. According to Lemma 1, if *X~GK(a, b, α, β)*, then $Y=\frac{1-(1-X^{a})^{b}}{(1-X^{a})^{b}}$~*Gamma* (*α*, *β*). In such a case for the random variable *Y*, the hazard rate function can be written as

$$\begin{aligned}\frac{1}{r(t)}&=\frac{1-F(t)}{f(t)}\\&=\frac{\int_t^{\infty}\frac{1}{\beta^{\alpha}\Gamma(\alpha)}w^{\alpha-1}\exp(-w/\beta)dw}{\frac{1}{\beta^{\alpha}\Gamma(\alpha)}t^{\alpha-1}\exp(-t/\%beta)}\\&=\int_t^{infty}\left(\tfrac{w}{t}\right)^{\alpha-1}\exp\left(-1/\beta(w-t)\right)dw\\&=\int_0^{\infty}\left(1+\tfrac{u}{t}\right)^{\alpha-1}\exp(-1/\beta u)du.\end{aligned} \tag{11}$$

Therefore, $r(t)=\left(\int_0^{\infty}\left(1+\frac{u}{t}\right)^{\alpha-1}\exp(-1/\beta u)du\right)^{-1}$. If $\alpha>1$, $\left(1+\frac{u}{t}\right)^{\alpha-1}$ is decreasing in t and hence *r(t)* is increasing, thereby and has a IFR. If $0<\alpha<1$, then

$\left(1+\frac{u}{t}\right)^{\alpha-1}$ is increasing in *t*, so *r(t)* decreases and hence has a DFR. Now, since *X* is a one-to-one function of *Y*, the hazard rate function of *X* will also follow the exact pattern.

Let *X* and *Y* be two random variables. *X* is said to be stochastically greater than or equal to *Y* denoted by $X\underset{st}{\geq}Y$ if $P(X>x)\geq P(Y>x)$ for all x in the support set of *X*.

Theorem 3. Suppose X~$GK(a_1,b_1,\alpha,\beta_1)$ and Y~$GK(a_2,b_2,\alpha,\beta_2)$. If $\beta_1>\beta_2$, $a_1>a_2$ and $b_1<b_2$. Then $X\underset{st}{\geq}Y$, for integer values of a_1 and a_2.

Proof. At first, we note that the incomplete gamma function $\Gamma(\alpha,x)$ is an increasing function of x for fixed α. For any real number $x\in(0,1)$, $\beta_1>\beta_2$, $a_1>a_2$, and $b_1<b_2$, we have

$$\beta_1^{-1}\left((1-x^{a_1})^{b_1}-1\right)\leq\beta_2^{-1}\left((1-x^{a_2})^{b_2}-1\right). \tag{12}$$

This implies that $\Gamma\left(\alpha,\beta_1^{-1}\left((1-x^{a_1})^{b_1}-1\right)\right)\leq\Gamma\left(\alpha,\beta_2^{-1}\left((1-x^{a_2})^{b_2}-1\right)\right)$. Equivalently, it implies that $P(X>x)\geq P(Y>x)$, and this completes the proof. W

Note: For fractional choices of a_1 and a_2, the reverse of the above inequality will hold.

4. Moments and mean deviations

For any $r \geq 1$,

$$E(X^r) = \int_0^1 x^r f(x)dx$$

$$= \frac{1}{\Gamma(\alpha)} \int_0^\infty \exp(-u) u^{\alpha-1} \left(1-(1+u\beta)^{-1/b}\right)^{r/a} du$$

$$\text{(on substitution } u = \frac{1-(1-x^a)^b}{\beta(1-x^a)^b})$$

$$= \frac{1}{\Gamma(\alpha)} \sum_{j=0}^{\infty} (-1)^j \binom{r/a}{j} \int_0^\infty \exp(-u) u^{\alpha-1} (1+u\beta)^{-j/b} du \tag{13}$$

$$= \frac{1}{\Gamma(\alpha)} \sum_{j=0}^{\infty}\sum_{k=0}^{\infty} (-1)^{j+k} \binom{r/a}{j} \binom{j/b+k-1}{k} \int_0^\infty \exp(-u) \beta^k u^{\alpha+k-1} du$$

$$= \frac{\beta^k}{\Gamma(\alpha)} \sum_{j=0}^{\infty}\sum_{k=0}^{\infty} (-1)^{j+k} \binom{r/a}{j} \binom{j/b+k-1}{k} \Gamma(\alpha+k).$$

Upper bounds for the r-th order moment: Since $\binom{n}{k} \leq \frac{n^k}{k!}$, for $1 \leq k \leq n$, from Eq. (13), one can write $E(X^r) \leq \left(\left(\frac{r}{a}\right)\left(\frac{j}{b}+k-1\right)\right) + \frac{\beta^k}{\Gamma(\alpha)} \sum_{j=0}^{\infty}\sum_{k=0}^{\infty} (-1)^{j+k} \frac{(r/a)^j}{j!} \left(\frac{(j/b+k-1)^k}{k!}\right) \Gamma(\alpha+k)$, provided r/a and $j/b+k-1$ are both integers. Employing successively, the generalized series expansion of $\left(1-(1+\beta u)^{-1/b}\right)^{j/a}$, the characteristic function for $X \sim GK(a,b,\alpha,\theta)$ will be given by [from Eq. (3)]

$$\phi_X(t) = \frac{1}{\Gamma(\alpha)} \int_0^1 e^{itx} f(x)dx$$

$$= \frac{1}{\Gamma(\alpha)} \int_0^\infty u^{\alpha-1} e^{-u} \exp\left(it\left(1-(1+\beta u)^{-1/b}\right)^{1/a}\right) du$$

$$\text{on substitution} \quad u = \frac{1-(1-x^a)^b}{\beta(1-x^a)^b} \tag{14}$$

$$= \frac{1}{\Gamma(\alpha)} \sum_{j=0}^{\infty} \int_0^\infty \frac{\left(it\left(1-(1+\beta u)^{-1/b}\right)^{1/a}\right)^j}{j!} u^{\alpha-1} e^{-u} du$$

$$= \frac{1}{\Gamma(\alpha)} \sum_{j=0}^{\infty}\sum_{k_1=0}^{\infty}\sum_{k_2=0}^{\infty} (-1)^{k_1+k_2} \beta^{k_2} (it)^j \binom{j/a}{k_1} \binom{k_1/b}{k_2} \Gamma(\alpha+k_2).$$

If j/a and k_1/b are integers then in Eq. (14), the second and third summations will stop at j/a and k_1/b, respectively.

If we denote the median by T, then the mean deviation from the mean, $D(\mu)$, and the mean deviation from the median, $D(T)$, can be written as

$$D(\mu) = E|X-\mu| = 2\mu G(\mu) - 2\int_{-\infty}^{\mu} xf(x)dx. \tag{15}$$

$$D(T) = E|X-T| = \mu - 2\int_{-\infty}^{T} xf(x)dx. \tag{16}$$

Now, consider

$$\begin{aligned} I_t &= \int_0^t xf(x)dx \\ &= \int_0^t xab \frac{\exp\left(-\frac{1-(1-x^a)^b}{\beta(1-x^a)^b}\right)}{\Gamma(\alpha)\beta^\alpha} \\ &\times \frac{1}{(1-x^a)^2}\left(\frac{1-(1-x^a)^b}{\beta(1-x^a)^b}\right)^{\alpha-1} x^{a-1}dx. \end{aligned} \tag{17}$$

Using the substitution $u = \frac{1-(1-x^a)^b}{\beta(1-x^a)^b}$ in Eq. (17), we obtain

$$\begin{aligned} I_t &= \frac{1}{\Gamma(\alpha)}\int_0^{\frac{1-(1-t^a)^b}{\beta(1-t^a)^b}} \left(1-(1+u\beta)^{-1/b}\right)^{1/a} u^{\alpha-1}e^{-u}du \\ &= \frac{1}{\Gamma(\alpha)}\sum_{j=0}^{\infty}\sum_{k=0}^{\infty}(-1)^j\beta^k(it)^j\binom{1/a}{j}\binom{j/b+k-1}{k}\Gamma(\alpha,\frac{1-(1-t^a)^b}{\beta(1-t^a)^b}), \end{aligned} \tag{18}$$

where we used successively binomial series expansion.

By using Eqs. (2) and (18), the mean deviation from the mean and the mean deviation from the median are, respectively, given by

$$\begin{aligned} D(\mu) &= 2\mu\frac{\Gamma\left(\alpha,\frac{1-(1-m^a)^b}{\beta(1-m^a)^b}\right)}{\Gamma(\alpha)} - 2I_\mu. \\ D(M) &= mu - 2I_M. \end{aligned} \tag{19}$$

4.1. Entropy

One useful measure of diversity for a probability model is given by Renyi's entropy. It is defined as $I_R(\rho) = (1-\rho)^{-1}\log\left(\int f^\rho(x)dx\right)$, where $\rho > 0$ and $\rho \neq 1$. If a random variable X has a GK distribution, then we have

$$f^\rho(x) = \left(\frac{ab}{\Gamma(\alpha)\beta^\alpha}\right)^\rho \exp\left(-\frac{\rho(1-(1-x^a)^b)}{\beta(1-x^a)^b}\right) \times \frac{1}{(1-x^a)^{2\rho}}\left(\frac{1-(1-x^a)^b}{\beta(1-x^a)^b}\right)^{\rho(\alpha-1)} x^{\rho(a-1)} \tag{20}$$

Next, consider the integral

$$\int_0^1 f^\rho(x)dx = \left(\tfrac{ab}{\Gamma(\alpha)\beta^\alpha}\right)^\rho \rho^{-1}\int_0^\infty u^{\alpha-1}\exp(-u)\left(1-(1+\beta u^{1/\rho})^{-1/b}\right)^{1/a}du$$
$$\text{(on substitution } u = \left(\tfrac{1-(1-x^a)^b}{\beta(1-x^a)^b}\right)^\rho\text{)}. \tag{21}$$

Now, using successive application of the generalized binomial expansion, we can write

$$\left(1-(1+\beta u^{1/\rho})^{-1/b}\right)^{1/a} = (-1)^{\rho-1}\sum_{j=0}^{\infty}\sum_{k=0}^{\infty}(-1)^j\binom{\rho-1}{j}\binom{1/b+k-1}{k}\beta^k u^{k/\rho}. \tag{22}$$

Hence, the integral in Eq. (21) reduces to

$$\int_0^1 f^\rho(x)dx = \left(\tfrac{ab}{\Gamma(\alpha)\beta^\alpha}\right)^\rho \rho^{\rho\alpha-k}(-1)^{\rho-1}\sum_{j=0}^{\infty}\sum_{k=0}^{\infty}(-1)^j\binom{\rho-1}{j}\binom{1/b+k-1}{k}\beta^k\Gamma(\rho\alpha+k) = \delta(\rho,\alpha,\beta,a,b),\ \text{say} \tag{23}$$

Therefore, the expression for the Renyi's entropy will be

$$I_R(\rho) = (1-\rho)^{-1}\log\left(\delta(\rho,\alpha,\beta,a,b)\right) \tag{24}$$

5. Maximum likelihood estimation

In this section, we address the parameter estimation of the GK(α, β, a, b) under the classical set up. Let X_1, X_2, ..., X_n be a random sample of size n drawn from the density Eq. (3). The log-likelihood function is given by

$$\begin{aligned}\ell = &-\alpha\log\beta - n\log\Gamma(\alpha) + n\log a + n\log b + (a-1)\sum_{i=1}^{n}\log X_i \\ &-\sum_{i=1}^{n}\frac{1-(1-X_i^a)^b}{\beta(1-X_i^a)^b} - 2\sum_{i=1}^{n}\log(1-X_i^a) + (\alpha-1)\sum_{i=1}^{n}\log\left(\frac{1-(1-X_i^a)^b}{\beta(1-X_i^a)^b}\right).\end{aligned} \tag{25}$$

The derivatives of Eq. (13) with respect to α, β, a, and b are given by

$$\frac{\partial}{\partial\alpha}\ell = -n\log\beta - \Psi(\alpha) + \sum_{i=1}^{n}\log\left(\frac{1-(1-X_i^a)^b}{\beta(1-X_i^a)^b}\right), \tag{26}$$

where $\Psi(\alpha) = \frac{\partial}{\partial\alpha}\log\Gamma(\alpha)$,

$$\frac{\partial}{\partial\beta}\ell = -\frac{\alpha}{\beta} + \beta^{-2}\sum_{i=1}^{n}\left(\frac{1-(1-X_i^a)^b}{\beta(1-X_i^a)^b} - (\alpha-1)\log\left(\frac{1-(1-X_i^a)^b}{\beta(1-X_i^a)^b}\right)\right). \tag{27}$$

$$\frac{\partial}{\partial a}\ell=\frac{n}{a}+\sum_{i=1}^{n}\log X_i+2\sum_{i=1}^{n}X_i^a(1-X_i^a)-1\log X_i$$
$$+\frac{b(\alpha-1)}{\beta}\sum_{i=1}^{n}\left(\frac{1-(1-X_i^a)^b}{\beta(1-X_i^a)^b}\right)^{-1}\frac{X_i^a\log X_i}{(1-X_i^a)^{b+1}}-\frac{1}{\beta}\sum_{i=1}^{n}\frac{X_i^a\log X_i}{(1-X_i^a)^{b+1}} \tag{28}$$

$$\frac{\partial}{\partial b}\ell=\frac{n}{b}+\frac{1}{\beta}\left(-1+\sum_{i=1}^{n}\log(1-X_i^a)\left(1-\left(\frac{\alpha-1}{\beta}\right)\frac{1-(1-X_i^a)^b}{(1-X_i^a)^b}\right)\right). \tag{29}$$

The MLEs $\hat{\alpha}$, $\hat{\beta}$, $\hat{a}$, and $\hat{b}$ are obtained by setting Eqs. (26–29) to zero and solving them simultaneously.

To estimate the model parameters, numerical iterative techniques must be used to solve these equations. We may investigate the global maxima of the log likelihood by setting different starting values for the parameters. The information matrix will be required for interval estimation. The elements of the 4 × 4 total observed information matrix (since expected values are difficult to calculate), $J(\vec{\theta})=J_{r,s}(\vec{\theta})$ (for $r,s=\alpha,\beta,a,b$), can be obtained from the authors under request, where $\vec{\theta}=(\alpha,\beta,a,b)$. The asymptotic distribution of $(\hat{\vec{\theta}}-\vec{\theta})$ is $N_4(\vec{0},K(\theta)^{-1})$, under the regularity conditions, where $K(\theta)=E\left(J(\vec{\theta})\right)$ is the expected information matrix, and $J(\hat{\vec{\theta}})^{-1}$ is the observed information matrix. The multivariate normal $N_4(\vec{0},K(\theta)^{-1})$ distribution can be used to construct approximate confidence intervals for the individual parameters.

5.1. Simulation study

In order to assess the performance of the MLEs, a small simulation study is performed using the statistical software R through the package (stats4), command MLE. The number of Monte Carlo replications was 20,000 For maximizing the log-likelihood function, we use the MaxBFGS subroutine with analytical derivatives. The evaluation of the estimates was performed based on the following quantities for each sample size; the empirical mean squared errors (MSEs) are calculated using the R package from the Monte Carlo replications. The MLEs are determined for each simulated data, say, $(\hat{\alpha}_i,\hat{\beta}_i,\hat{a}_i,\hat{b}_i)$ for $i=1,2,\ldots,20,000$, and the biases and MSEs are computed by

$$bias_h(n)=\frac{1}{20000}\sum_{i=1}^{20000}(\hat{h}_i-h), \tag{30}$$

and

$$MSE_h(n)=\frac{1}{20000}\sum_{i=1}^{20000}(\hat{h}_i-h)^2, \tag{31}$$

for $h=\alpha,\beta,a,b$. We consider the sample sizes at n = 100, 200, and 500 and consider different values for the parameters . The empirical results are given in **Table 1**. The figures in **Table 1** indicate that the estimates are quite stable and, more importantly, are close to the true values for these sample sizes. Furthermore, as the sample size increases, the MSEs decrease as expected.

Sample size	Actual value				Bias				MSE			
n	α	β	a	b	$\hat{\alpha}$	$\hat{\beta}$	$\hat{a}$	$\hat{b}$	$\hat{\alpha}$	$\hat{\beta}$	$\hat{a}$	$\hat{b}$
100	0.5	0.5	2	4	−0.417	−0.419	0.355	−0.393	0.051	0.046	0.053	0.97
	0.5	0.5	3	5	−0.773	0.324	−0.214	−0.342	0.018	0.042	0.098	0.626
	0.7	0.8	4	3	0.489	−0.246	−0.623	0.482	0.015	0.121	0.106	0.167
	0.9	0.7	6	4	0.188	0.979	−0.509	0.056	0.048	0.022	0.044	0.114
	1	1.5	0.9	0.6	0.178	−0.498	−0.429	−0.545	0.427	0.028	0.092	0.495
	1.5	2	0.6	0.8	−0.084	−0.363	−0.405	−0.220	0.953	0.018	0.073	0.572
200	0.5	0.5	2	4	.072	0.361	0.049	0.073	0.022	0.023	0.024	0.313
	0.5	0.5	3	5	0.518	0.184	0.084	0.115	0.008	0.022	0.045	0.578
	0.7	0.8	4	3	0.316	0.159	0.050	−0.329	0.006	0.059	0.044	0.158
	0.9	0.7	6	4	0.137	−0.049	0.131	−0.032	0.018	0.010	0.020	0.095
	1	1.5	0.9	0.6	0.125	0.475	0.086	0.242	0.064	0.013	0.028	0.147
	1.5	2	0.6	0.8	0.034	0.173	0.224	−0.150	0.401	0.008	0.036	0.432
500	0.5	0.5	2	4	−0.046	−0.028	−0.036	−0.047	0.009	0.011	0.011	0.084
	0.5	0.5	3	5	−0.051	−0.111	−0.035	0.002	0.004	0.010	0.022	0.018
	0.7	0.8	4	3	−0.0730	−0.052	0.046	−0.022	0.004	0.031	0.024	0.036
	0.9	0.7	6	4	−0.102	−0.02	−0.098	−0.023	0.008	0.005	0.010	0.021
	1	1.5	0.9	0.6	−0.078	−0.052	−0.017	0.003	0.027	0.007	0.012	0.015
	1.5	2	0.6	0.8	0.007	−0.069	0.066	−0.085	0.136	0.004	0.015	0.013

Table 1. Bias and MSE of the estimates under the maximum likelihood method.

6. Reliability parameter

The reliability parameter R is defined as $R = P(X > Y)$, where X and Y are independent random variables. For a detailed study on the possible applications of the reliability parameter, an interested reader is suggested to look at Ref. [4, 5]. If X and Y are two continuous and independent random variables with the cdf's $F_1(x)$ and $F_2(y)$ and their pdf's $f_1(x)$ and $f_2(y)$, respectively, then the reliability parameter R can be written as

$$R = P(X > Y) = \int_{-\infty}^{\infty} F_2(t) f_1(t) dt. \tag{32}$$

Theorem 4. Let X~GK(a, b, α_1, β_1) and Y~(a, b, α_2, β_2), then

$$R = \sum_{p=0}^{\infty} \frac{(-1)^p}{p!(\alpha_2 + p)\Gamma(\alpha_1)} \left(\frac{\beta_1}{\beta_2}\right)^{p+\alpha_2} \Gamma(\alpha_1 + \alpha_2 + p). \tag{33}$$

Proof: From Eqs. (2) and (3), we have

$$R=\int_0^1 \gamma_1\left(\alpha_2,\frac{1-(1-t^a)^b}{\beta_2(1-t^a)^b}\right)\frac{ab\exp\left(-\frac{1-(1-t^a)^b}{\beta_1(1-t^a)^b}\right)}{\Gamma(\alpha_1)\beta_1^{\alpha}}\times\frac{1}{(1-t^a)^2}\left(\frac{1-(1-t^a)^b}{\beta_1(1-t^a)^b}\right)^{\alpha_1-1}t^{a-1}dt. \quad (34)$$

Using the series expansion for the incomplete gamma function $\gamma_1(k,x)=x^k\sum_{p=0}^{\infty}\frac{(-x)^p}{k!(k+p)}$, and using the substitution $u=\frac{1-(1-t^a)^b}{\beta_1(1-t^a)^b}$, Eq. (34) reduces to

$$\begin{aligned} R&=\sum_{p=0}^{\infty}\frac{(-1)^p}{p!(\alpha_2+p)\Gamma(\alpha_1)}\left(\frac{u\beta_1}{\beta_2}\right)^{p+\alpha_2}u^{\alpha_1-1}\exp(-u)du\\ &=\sum_{p=0}^{\infty}\frac{(-1)^p}{p!(\alpha_2+p)\Gamma(\alpha_1)}\left(\frac{\beta_1}{\beta_2}\right)^{p+\alpha_2}\Gamma(\alpha_1+\alpha_2+p). \end{aligned} \quad (35)$$

Hence the proof. W

7. Order statistics

Here, we derive the general r-th order statistic and the large sample distribution of the sample minimum and the sample maximum based on a random sample of size n from the GK(α, β, a, b) distribution. The corresponding density function of the r-th order statistic, $X_{r:n}$, from Eq. (3) will be

$$\begin{aligned} f_{X_{r:n}}(x)&=\frac{1}{B(r,n-r+1)}(F(x))^{r-1}(1-F(x))^{n-r}f(x)\\ &=\frac{f(x)}{B(r,n-r+1)}\sum_{j=0}^{r-1}(-1)^j\binom{r-1}{j}\left(\frac{\Gamma\left(\alpha,\beta^{-1}\frac{1-(1-x_{r:n}^a)^b}{(1-x_{r:n}^a)^b}\right)}{\Gamma(\alpha)}\right)^{n-r+j}\times I(0<x<1). \end{aligned} \quad (36)$$

Using the series expression for the incomplete gamma function: $\gamma_1(\alpha,x)=\sum_{k=0}^{\infty}\frac{e^{-x}(x)^{\alpha+k}}{\alpha(\alpha+1)\cdots(\alpha+k)}$, the pdf of $X_{r:n}$ can be written as

$$\begin{aligned} f_{r:n}(x)&=\frac{1}{B(r,n-r+1)}f(x)\sum_{j=0}^{r-1}(-1)^j\binom{r-1}{j}\left(\sum_{k=0}^{\infty}\frac{\exp\left(-\frac{1-(1-x_{r:n}^a)^b}{\beta(1-x_{r:n}^a)^b}\right)\left(\frac{1-(1-x_{r:n}^a)^b}{\beta(1-x_{r:n}^a)^b}\right)^{\alpha+k}}{\Gamma(\alpha)\alpha(\alpha+1)\cdots(\alpha+k)}\right)^{n-r+j}\\ &=\frac{f(x)}{B(r,n-r+1)}\sum_{j=0}^{r-1}\sum_{k_1}^{\infty}\cdots\sum_{k_{n-r+j}=0}^{\infty}(-1)^{j+s_k}\binom{r-1}{j}\exp\left(-(n-r+j)\frac{1-(1-x_{r:n}^a)^b}{\beta(1-x_{r:n}^a)^b}\right)\\ &\times\frac{\left(\frac{1-(1-x_{r:n}^a)^b}{(1-x_{r:n}^a)^b}\right)^{s_k+(n-r+j)\alpha}}{(\Gamma(\alpha))^{n-r+j}\beta^{s_k+(n-r+j)\alpha}p_k}\\ &=\frac{1}{B(r,n-r+1)}\sum_{j=0}^{n-r}\sum_{k_1}^{\infty}\cdots\sum_{k_{n-r+j}=0}^{\infty}(-1)^{j+s_k}\binom{r-1}{j}\\ &\times\frac{\Gamma(s_k+(r+j)\alpha)}{(\Gamma(\alpha))^{n-r+j}p_k}f(x|s_k+(n-r+j)\alpha,\beta,a,b), \end{aligned} \quad (37)$$

where $s_k = \sum_{i=1}^{n-r+j} k_i$ and $p_k = \prod_{i=1}^{n-r+j} (k_i + \alpha)$.

From Eq. (37), it is interesting to note that the pdf of the r-th order statistic $X_{r:n}$ can be expressed as an infinite sum of the GK pdf 's.

8. Application

Here, we consider two well-known illustrative data sets which are used to show the efficacy of the GK distribution. For details on these two data sets [6, 7], the second data set in **Table 2** is from Ref. [8], and it represents the fatigue life of 6061-T6 aluminum coupons cut parallel with the direction of rolling and oscillated at 18 cycles per second. The GK distribution is fitted to the first data set and compared the result with the Kumaraswamy, gamma-uniform [9], and beta-Pareto [10]. These results are reported in **Table 3**. The results show that gamma-uniform, GK distributions provide adequate fit to the data. **Figure 3** displays the empirical and the fitted cumulative distribution functions. This figure supports the results in **Table 3**. A close look at **Figure 3** indicates that GK distribution provides better fit to the left tail than the gamma-uniform distribution. This is due to the fact that GK distribution can have longer left tail (**Figure 3**).

70	90	96	97	99	100	103	104	104	105
107	108	108	108	109	109	112	112	113	114
114	114	116	119	120	120	120	121	121	123
124	124	124	124	124	128	128	129	129	130
130	130	131	131	131	131	131	132	132	132
133	134	134	134	134	134	136	136	137	138
142	142	144	144	145	146	148	148	149	151
151	152	155	156	157	157	157	157	158	159
162	163	163	164	166	166	168	170	174	196
212									

Table 2. Fatigue life of 6061-T6 aluminum data.

In addition, to check the goodness-of-fit of all statistical models, several other goodness-of-fit statistics are used and are computed using computational package Mathematica. The MLEs are computed using N maximize technique as well as the measures of goodness-of-fit statistics including the log-likelihood function evaluated at the MLEs (l), Akaike information criterion (AIC), corrected Akaike information criterion (AICC), consistent Akaike information criterion (CAIC), the Anderson-Darling (A^*), the Cramer-von Mises (W^*), and the Kolmogrov-Smirnov (K-S) statistics with their p values to compare the fitted models. These statistics are used to evaluate how closely a specific distribution with cdf (2) fits the corresponding empirical distribution for a given data set. The distribution with better fit than the others will be the one having the smallest statistics and largest p value. Alternatively, the distribution for which one obtains the smallest of each of these criteria (i.e., AIC, AICC, K-S, etc.) will be most suitable one. The mathematical equations of those statistics are given by

- $AIC = -2\ell(\hat{\theta}) + 2q$
- $AICC = AIC + \frac{2q(q+1)}{n-q-1}$
- $CAIC = -2\ell(\hat{\theta}) + \frac{2qn}{n-q-1}$
- $A_0^* = \left(\frac{2.25}{n^2} + \frac{0.75}{n} + 1\right)\left(-n - \frac{1}{n}\sum_{i=1}^{n}(2i-1)\log\left(z_i(1-z_{n-i+1})\right)\right)$
- $W_0^* = \left(\frac{0.5}{n} + 1\right)\left[\left(z_i - \frac{2i-1}{2n}\right)^2 + \frac{1}{12n}\right]$
- $K\text{-}S = Max\left(\frac{i}{n} - z_i, z_i - \frac{i-1}{n}\right),$

Distribution	Kumaraswamy	Gamma-uniform	Beta-Pareto	Gamma-Kumaraswamy
Parameter estimates	$\hat{a} = 0.653$	$\hat{\alpha} = 7.528$	$\hat{c} = 5.048$	$\hat{\alpha} = 7.891$
	$\hat{b} = 1.1182$	$\hat{\beta} = 2.731$	$\hat{\beta} = 0.401$	$\hat{\beta} = 0.785$
		$\hat{a} = 6.49$	$\hat{\theta} = 6.417$	$\hat{a} = 5.352$
		$\hat{b} = 0.932$		$\hat{b} = 1.735$
Log likelihood	−162.34	−116.58	−113.36	−113.25
AIC	217.38	119.45	167.78	107.85
AICC	218.36	120.37	168.91	108.43
CAIC	218.36	120.37	168.91	108.43
A_0^*	12.164	0.5951	1.3125	0.4282
W_0^*	2.8937	0.0931	0.1317	0.04893
K-S	0.5290	0.09521	0.4245	0.0492
K-S p-value	0.0000	0.9140	0.8374	0.9978

Table 3. Goodness of fit of deep-groove ball bearings data.

where $\ell(\hat{\theta})$ denotes the log-likelihood function evaluated at the maximum likelihood estimates, q is the number of parameters, n is the sample size and $z_i = cdf(y_i)$, the y_i 's being the ordered observations.

Lieblein and Zelen [6] proposed a five parameter beta generalized Pareto distribution and fitted the data in **Table 4** and compared the result with beta-Pareto and other known distributions. The results of fitting beta generalized Pareto and beta-Pareto from Ref. [8] are reported in **Table 4** along with the results of fitting the Pareto (IV) and GK distributions to the data. The KS value from **Table 4** indicates that the GK distribution provides the best fit. The fact that GK distribution has the least number of parameters than beta generalized Pareto and beta-Pareto adds an extra advantage over them. **Figure 4** displays the empirical and the fitted cumulative distribution functions. This figure supports the results in **Table 4**.

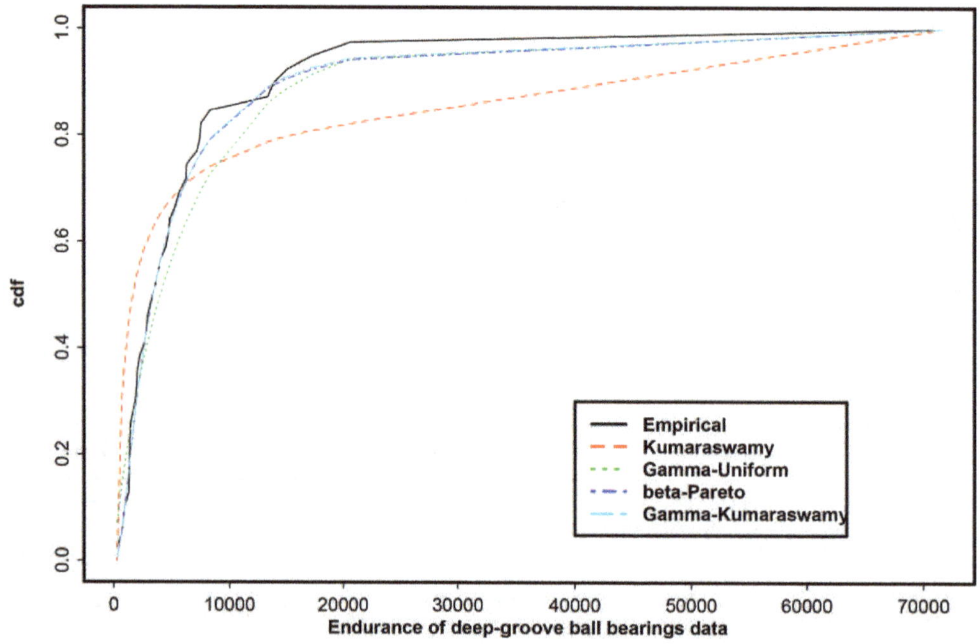

Figure 3. cdf for fitted distributions of the endurance of deep-groove ball bearings data.

Distribution	Kumaraswamy	Beta-Pareto	Beta generalized Pareto	gamma-Kumaraswamy
Parameter estimates	$\hat{\delta} = 0.235$	$\hat{\alpha} = 485.470$	$\hat{\alpha} = 12.112$	$\hat{\alpha} = 4.87$
	$\hat{\gamma} = 2.4926$	$\hat{\beta} = 162.060$	$\hat{\beta} = 1.702$	$\hat{\beta} = 3.352$
		$\hat{k} = 0.3943$	$\hat{\mu} = 40.564$	$\hat{a} = 1.1722$
		$\hat{\theta} = 3.910$	$\hat{k} = 0.273$	$\hat{b} = 2.0154$
			$\hat{\theta} = 54.837$	
Log likelihood	–572.39	–517.33	–457.85	–417.36
AIC	1018.83	925.30	925.70	878.45
AICC	1018.956	926.01	926.84	879.63
CAIC	1018.956	926.01	926.84	879.63
A_0^*	4.1745	1.0083	0.6584	0.4921
W_0^*	0.6739	0.2827	0.1195	0.0822
K-S	0.4723	0.091	0.070	0.064
K-S p-value	0.000	0.248	0.537	0.736

Table 4. Parameter estimates for the fatigue life of 6061-T6 aluminum coupons data.

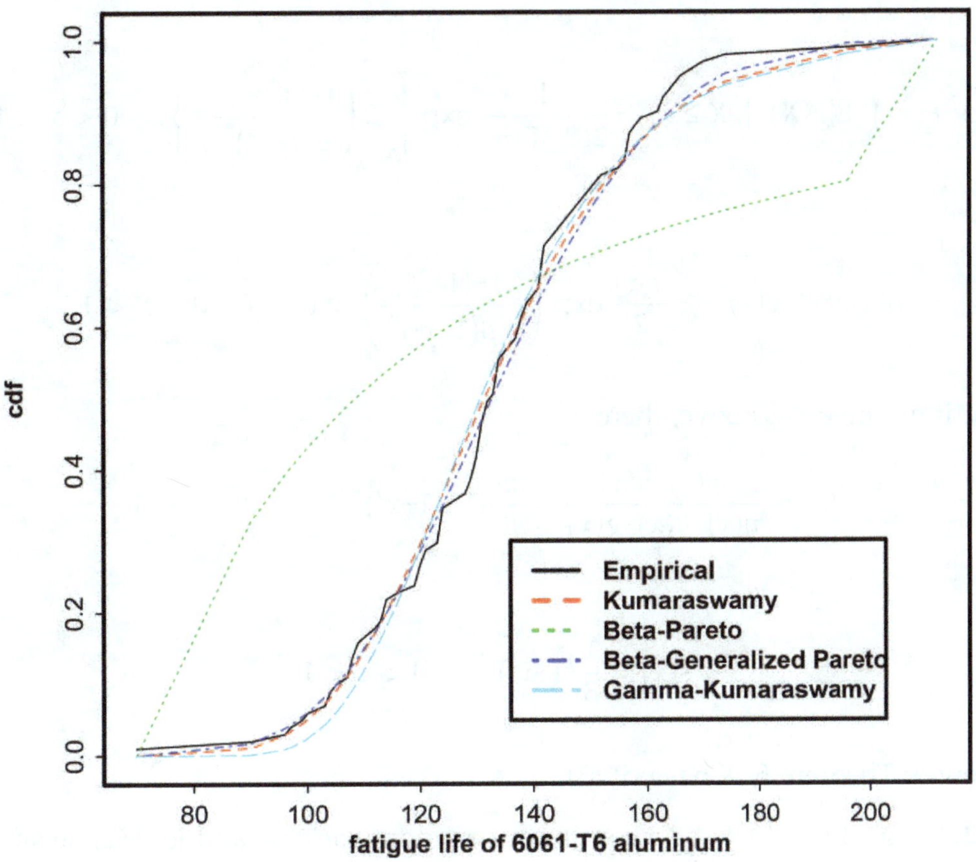

Figure 4. cdf for fitted distributions of the fatigue life of 6061-T6 Aluminum data.

9. Characterization of GK distribution

In this section, we present characterizations of GK distribution in terms of the ratio of two truncated moments. For the previous works done in this direction, we refer the interested readers to Glänzel [11–14] and Hamedani [15–17]. For our characterization results, we employ a theorem due to Ref. [11], see for further details. The advantage of the characterizations given here is that *cdf F* need not have a closed form. We present here a corollary as a direct application of the theorem discussed in details in Ref. [11].

Corollary 1. Let $X:\Omega\to(0,1)$ be a continuous random variable and let $h(x)=\beta^{\alpha-1}(1-x^a)^{b(\alpha-2)+1}[1-(1-x^a)^b]^{1-\alpha}$ and $g(x)=h(x)\exp\left(-\frac{1-(1-x^a)^b}{\beta(1-x^a)^b}\right)$ for $x\in(0,1)$. Then X has pdf (3) if and only if the function η defined in Theorem 5 has the form

$$\eta(x)=\frac{1}{2}\exp\left(-\frac{1-(1-x^a)^b}{\beta(1-x^a)^b}\right),\quad 0<x<1. \tag{38}$$

Proof. Let X has pdf (3), then

$$\left(1-F(x)\right)\quad \mathbf{E}[h(X)\mid X\geq x]=\frac{1}{\beta^{\alpha-1}\Gamma(\alpha)}\exp\left(-\frac{1-(1-x^a)^b}{\beta(1-x^a)^b}\right),\quad 0<x<1, \tag{39}$$

and

$$\left(1\text{-}F(x)\right)\ \mathbf{E}[g(X)\ |\ X \geq x] = \frac{1}{2\beta^{\alpha-1}\Gamma(\alpha)} \exp\left\{-2\left(\frac{1-(1-x^a)^b}{\beta(1-x^a)^b}\right)\right\}, \quad 0 < x < 1, \tag{40}$$

and finally

$$\eta(x)h(x)\text{-}g(x) = -\frac{h(x)}{2} \exp\left(-\frac{1-(1-x^a)^b}{\beta(1-x^a)^b}\right) < 0, \quad for\ 0 < x < 1. \tag{41}$$

Conversely, if η is given as above, then

$$s'(x) = \frac{\eta'(x)\ h(x)}{\eta(x)\ h(x)\text{-}g(x)} = \frac{ab}{\beta}x^{a-1}(1-x^a)^{-(b+1)}, \quad 0 < x < 1, \tag{42}$$

and hence

$$s(x) = \frac{1}{\beta}(1-x^a)^{-b}, \quad 0 < x < 1. \tag{43}$$

Now, in view of Theorem 5, X has pdf (3).

Corollary 2. Let $X : \Omega \to (0, 1)$ be a continuous random variable and let $h(x)$ be as in Proposition 1. Then, X has pdf (3) if and only if there exist functions g and η defined in Theorem 5 satisfying the differential equation

$$\frac{\eta'(x)h(x)}{\eta(x)h(x)\text{-}g(x)} = \frac{ab}{\beta}x^{a-1}(1-x^a)^{-(b+1)}, \quad 0 < x < 1. \tag{44}$$

Remarks 1. (*a*) The general solution of the differential equation in Corollary 1 is

$$\eta(x) = \exp\left(\frac{1-(1-x^a)^b}{\beta(1-x^a)^b}\right)\left[-\int\frac{ab}{\beta}x^{a-1}(1-x^a)^{-(b+1)}\exp\left(-\frac{1-(1-x^a)^b}{\beta(1-x^a)^b}\right)\left(h(x)\right)^{-1}g(x)dx + D\right], \tag{45}$$

for $0 < x < 1$, where D is a constant. One set of appropriate functions is given in Proposition 1 with $D = 0$

(*b*) Clearly, there are other triplets of functions (h, g, η) satisfying the conditions of Theorem 5. We presented one such triplet in Proposition 1.

10. Concluding remarks

A special case of the gamma-generated family of distributions, the gamma-Kumaraswamy distribution, is defined and studied. Various properties of the gamma-Kumaraswamy distribution are investigated, including moments, hazard function, and reliability parameter. The new model includes as special sub-models the gamma and Kumaraswamy distribution. Also, we provide various characterizations of the gamma-Kumaraswamy distribution. An application to a real data set shows that the fit of the new model is superior to the fits of its main sub-models. As future work related to this univariate GK model, we will consider the following:

- A natural bivariate extension to the model in Eq. (1) would be

$$f(x,y) \propto \frac{g(x,y)}{\Gamma(\alpha)\beta^{\alpha}\overline{G}^{2}(x,y)} \exp\left(-\frac{g(x,y)}{\beta barG(x,y)}\right)\left(\frac{g(x,y)}{\beta barG(x,y)}\right)^{\alpha-1}, \quad x>0, y>0. \tag{46}$$

  In this case, exact evaluation of the normalizing constant would be difficult to obtain, even for a simple analytic expression of a baseline bivariate distribution function, $G(x, y)$. Numerical methods such as Monte Carlo methods of integration might be useful here. We will study and discuss structural properties of such a bivariate GK model.

- Extension of the proposed univariate GK model to multivariate GK models and discuss the associated inferential issues. It is noteworthy to mention that classical methods of estimation, such as for example, maximum likelihood method of estimation might not be a good strategy because of the enormous number of model parameters. An appropriate Bayesian inference might be the only remedy. In that case, we will separately study two different cases of estimation: (a) with non-informative priors and (b) with full conditional conjugate priors (Gibbs sampling). Since the GK distribution is in the one parameter exponential family, a reasonable choice for priors for α and β might well be gamma priors with appropriate choice of hyper-parameters. For prior choices of the parameters that are from the baseline $G(.)$ distribution function, a data-driven prior approach will be more suitable.

- A discrete analog of the univariate GK model with a possible application in modeling rare events.

- Construction of a new class of GK mixture models by adopting Marshall-Olkin method of obtaining new distribution.

Author details

Indranil Ghosh[1]* and Gholamhossein G. Hamedani[2]

*Address all correspondence to: jamesbond.indranil@gmail.com

1 Department of Mathematics and Statistics, University of North Carolina, Wilmington, Wilmington, NC, USA

2 Department of Mathematics, Statistics and Computer Science, Marquette University, Milwaukee, WI, USA

References

[1] Johnson, N.L., Kotz, S., and Balakrishnan, N.: Continuous Univariate Distributions. John Wiley and Sons; New York, 1994. Vol. 2.

[2] Kumaraswamy, P.: Generalized probability density-function for double-bounded random-processes. Journal of Hydrology, 1980; **462**: 79–88.

[3] Zografos, K., and Balakrishnan, N.: On families of beta and generalized gamma-generated distributions and associated inference. Statistical Methodology, 2009; **4**: 344–362.

[4] Hall, I.J.: Approximate one-sided tolerance limits for the difference or sum of two independent normal variates. Journal of Qualitative Technology, 1984; **16**: 15–19.

[5] Weerahandi, S., and Johnson, R.A.: Testing reliability in a stress-strength model when X and Y are normally distributed. Technometrics, 1992; **38**: 83–91.

[6] Lieblein, J., and Zelen, M.: Statistical investigation of the fatigue life of deep-groove ball bearings. Journal of Research of the National Bureau of Standards, 1956; **57**: 273–316.

[7] Alzaatreh, A., and Ghosh, I.: A study of the Gamma-Pareto (IV) distribution and its applications. Communications in Statistics – Theory and Methods. 2016;**45**: 636–654, doi:10.1080/03610926. 2013.834453

[8] Mahmoudi, E.: The beta generalized Pareto distribution with application to lifetime data. Mathematics and Computers in Simulation, 2011; **81**: 2414–2430.

[9] Torabi, H., and Hedesh, N.M.: The gamma-uniform distribution and its applications. Kybernetika-Praha, 2011; **48**: 16–30.

[10] Akinsete, A., Famoye, F., and Lee, C.: The beta-Pareto distribution. Statistics, 2008; **42**: 547–563.

[11] Glänzel, W.: A characterization theorem based on truncated moments and its application to some distribution families. Mathematical Statistics and Probability Theory, 1987: 75–84.

[12] Glänzel, W.: Some consequences of a characterization theorem based on truncated moments. Statistics, 1990; **21**: 613–618.

[13] Glänzel, W., Telcs, A., and Schubert, A.: Characterization by truncated moments and its application to Pearson-type distributions. Zeitschrift fur Wahrscheinlichkeitstheorie und verwandte Gebiete, 1984; **66**: 173–183.

[14] Glänzel, W., and Hamedani, G.G.: Characterizations of univariate continuous distributions. Studia Scientiarum Mathematicarum Hungarica, 2001; **37**: 83–118.

[15] Hamedani, G.G.: Characterizations of univariate continuous distributions. Studia Scientiarum Mathematicarum Hungarica, 2002; **39**: 407–424.

[16] Hamedani, G.G.: Characterizations of univariate continuous distributions. Studia Scientiarum Mathematicarum Hungarica, 2006; **43**: 361–385.

[17] Hamedani, G.G.: Characterizations of continuous univariate distributions based on the truncated moments of functions of order statistics. Studia Scientiarum Mathematicarum Hungarica, 2010; **47**: 462–484.

5

A Practical Approach to Evaluating the Economic and Technical Feasibility of LED Luminaires

Sean Schmidt and Suzanna Long

Abstract

LED roadway luminaires are currently under consideration for widespread implementation with departments of transportation, facilities managers, and city planners. This research focuses on a case study in Missouri and presents relevant research findings calculated by the authors as part of a project funded by the Missouri Department of Transportation. Although high-pressure sodium (HPS) luminaires have been the standard product for roadway illumination, advances in LED technologies have led many departments of transportation to consider them as viable options along state routes. For this case study, pilot sites were developed across the state of Missouri in sites assessed as moderately busy, medium pedestrian conflict zones. These zones were along roadways with an R3 pavement classification. This case study details the economic feasibility findings from the study; a life cycle cost approach was used. In addition, a technical feasibility analysis was conducted to determine fit with Illumination Engineering Society (IES) standards for the traffic pattern and pavement classification at study sites. Key findings reveal that LED roadway luminaires fail to outperform HPS in their current design, but may become technically and economically feasible in the future.

Keywords: LED roadway luminaires, life cycle cost evaluation, field data, energy consumption, environmental impacts

1. Introduction

As high-pressure sodium (HPS) roadway luminaires reach the end of their product life cycle, many states and local agencies, as well as city planners and utilities are considering LED roadway luminaires as a replacement product [1]. Manufacturers of LED luminaires promote their benefits as longer useful life, reduced operations and maintenance costs, reduced environmental impact, and reduced energy cost. This case study presents a quantitative method for assessing technical and economic feasibility of LED roadway luminaires for those considering the product for widespread implementation.

Previously, research has been completed on LED luminaires in the field case studies sponsored by the Department of Energy's Energy Efficiency and Renewable Energy (EERE) program [2]. Research has also shown the life cycle costs of LEDs may be higher than high-pressure sodium luminaires for both collector and local roads [3, 4]. The Illumination Engineering Society (IES) of North America standards for roadway illumination vary depending on three factors: classification of roadway, pedestrian conflict potential, and pavement classification. For example, if a roadway was classified as a moderately traveled major route with a low pedestrian conflict potential and an R3 pavement classification, then the minimum maintained average illumination would be 9.0 lux [5]. If the same roadway's pedestrian conflict potential was reclassified to be a high potential of pedestrian conflict, then the minimum maintained average illuminance increases to 17.0 lux. The Illinois Center for Transportation produced a report providing background information on LED luminaires [6]. This report covers optics of LED lighting, advantages and drawbacks of LED lighting, and summarizes work previously completed by several GATEWAY demonstrations.

Manufacturers, government agencies, and utilities are collaborating to produce effective LED roadway luminaires. One such effort is the Department of Energy's Solid-State lighting GATEWAY Demonstration programs, which investigate real-world application of solid-state lighting technologies in various fields, such as roadway illumination, sidewalk illumination, and parking lot illumination. These programs have performed feasibility analyses on several types of LED luminaires across several uses. Thus far, the program has published reports on the use of LED lighting in parking lot [7] and minor roadway lighting [8]. Research has previously been performed on combining an economic analysis with a product performance analysis to develop street lighting standards [9–12]. In addition, the benefits of LEDs have been investigated in a previous research, such as reduced concern of the power factor of electricity loads [13]. Another potential benefit is the ability to rapidly start luminaires without harmful impact on the luminaire's lifetime, which can allow for "smarter" usage of lighting systems [14].

The Missouri Department of Transportation commissioned research into the feasibility of replacing existing high-intensity-discharge roadway lighting with LED lighting luminaires. The goals of this research include a technical analysis on the ability of LED luminaires to meet the minimum performance standards set by the Illumination Engineering Society and an economic analysis to compare the life cycle costs of replacing existing lighting luminaires with LED luminaires.

2. Evaluation of led roadway luminaires

2.1. LED luminaire data collection methodology

Illumination readings were collected from LED luminaire testing sites throughout the state of Missouri. The luminaires studied are currently used on roadways throughout Missouri. These readings were collected for LEDs produced by several manufacturers at three HPS equivalent power ratings: 150, 250, and 400 W. A total of eight unique manufacturer's LED luminaires were studied in this research.

Data collection points are based on a function of the pole spacing between luminaires and the width of the traffic lane at the location of the luminaire. Using intervals of one quarter of the distance between the target pole and adjacent poles minimizes interference caused by nearby streetlights. The pole spacing, roadway width, the distance between the pole and the outer lane, and the location of the luminaire were measured, in feet, for each luminaire using a perambulator. In order to minimize the impact of nearby sources of light, illuminance readings were collected such that the readings were directed toward the target luminaire. An illuminance meter was used to measure the lux for each field data location. The illuminance meter is greatly impacted by the direction in which the eyelet of the device points. Therefore, in order to minimize error, the maximum reading was recorded for each data point. Data was collected in an interval based on illumination pole spacing, or the distance between two luminaires. Data was collected in intervals of ¼ of the pole spacing. Pole spacing varied between data collection sites. Perpendicular data collection intervals along the road were collected in intervals equal to one lane of traffic, which in most cases was approximately 12 feet (3.66 m).

Key characteristics for the field study are repeated from the final report of the funded MoDOT project [15]. For each luminaire, 31 readings, including 15 readings at the ground level and 15 readings elevated 18 inches above the ground level, were collected. In addition, field data includes one ambient reading collected from a nonilluminated area. Ambient illuminance readings were collected approximately 20 feet behind the luminaire in order to be outside of the illuminated area. In order to determine the role of naturally occurring light sources, ambient readings were collected. This includes ambient lighting from nearby outdoor area lighting. To calculate adjusted field readings, ambient readings were subtracted from the field readings. This was then compared to the .ies file data for each studied luminaire. **Figure 1**

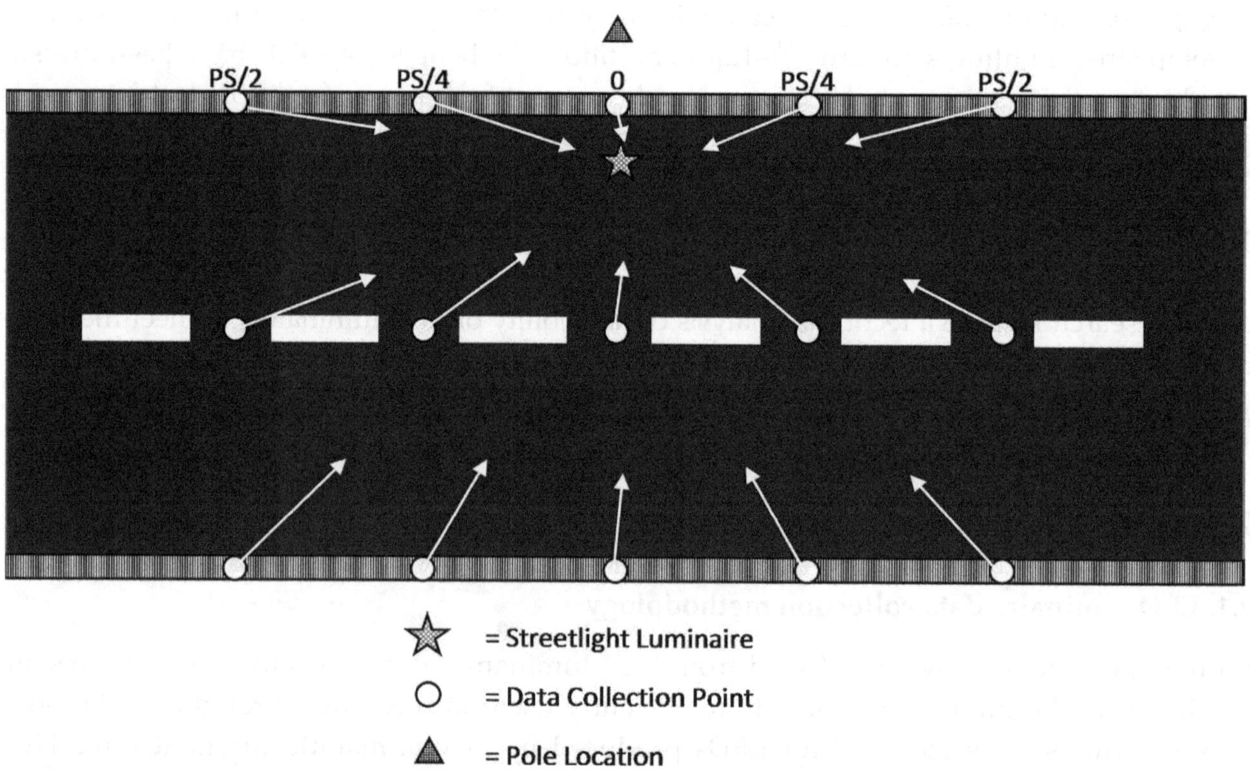

Figure 1. LED field testing methodology.

	Standard	LED A		LED B		LED C		LED D		LED E		LED F		LED G		LED H	
		Field data	IES data	Field data	IES data	Field data	IES data	Field data	IES data	Field data	IES data	Field data	IES data	Field data	IES data	Field data	IES data
Max (lux)		12.8	12.1	32.7	25.3	33.5	49.0	8.9	9.4	30.5	30.0	38.6	44.6	39.0	43.4	35.1	41.4
Min (lux)		4.2	3.2	8.0	2.4	4.0	2.5	2.0	2.4	7.1	6.1	9.8	4.4	2.0	2.5	4.4	2.1
Avg (lux)	>13	7.2	7.2	20.1	12.0	11.6	9.4	5.6	4.2	16.5	14.8	18.8	14.7	18.0	16.0	17.5	17.7
Avg/min	<3	1.7	2.2	2.5	5.0	2.9	3.8	2.8	1.8	2.3	2.4	1.9	3.3	8.9	6.4	4.0	8.4

Table 1. LED field data and manufacturer claims.

indicates the locations used for data collection points as well as the direction of the illuminance meter.

Once the field data collection phase ended, the manufacturer's .ies file for each luminaire was compared with field results to validate the manufacturer's claims. The variation between the field data and each manufacturer's claim was analyzed and is shown in figures within the field data evaluation and assessment section. Standards were created by the Illumination Engineering Society and are set in RP-08 [5]. These IES standards set a minimum of 13.0 lux for moderately busy, medium pedestrian conflict roads with R3 pavement classification. The desired average:minimum uniformity ratio for such a road is 3.0. Using the previously mentioned methodology, field data was collected for eight different LED luminaires across three HPS equivalent power ratings (150, 250, and 400 W). Five 150, two 250, and one 400 W equivalent luminaires were studied in this research. All field data collected was then compared to the IES standards provided by the manufacturer and the average to minimum uniformity ratios were calculated in accordance to the IES RP-08 publication [5]. All tested luminaires were installed with 9 months of data collection, therefore light loss factors were not applied to collected illuminance values.

Four out of the eight luminaires met the minimum average illuminance criteria of 13.0 lux at 30 foot mounting heights for both field readings and manufacturer's claims. The field readings for LED B fulfilled the criteria for minimum average illuminance and average to minimum ratio, however, the manufacturer's provided .ies file did not meet this claim, therefore LED B is deemed as technically infeasible. LEDs A, C, and D were deemed technically infeasible due to a combination of field readings and/or manufacturer's claims. LEDs F, G, and H were deemed technically infeasible due to their average to minimum ratios exceeding the 3.0 recommendation set by IES. The technical feasibility of HPS luminaires was not evaluated due to their current use in the transportation lighting field. LED fixtures were tested at mounting heights of 30 foot or less, since a significant portion of luminaires in Missouri is installed at these heights (**Table 1**).

3. Economic feasibility analysis

In order to conduct a thorough economic feasibility analysis of LED luminaires, several factors must be considered. These factors were originally reported in the MoDOT report and are repeated here [15]. Equivalencies are determined by grouping luminaires for comparison with the most appropriate high-pressure sodium luminaire. Manufacturers have worked to produce LED luminaires that are specifically designed as equivalent replacements for traditional high-intensity-discharge (HID) lamps. This allows transportation organizations the option of directly replacing traditional luminaires with LED luminaires, but other factors must be considered as well.

Second, the fiscal feasibility of LED luminaires rely heavily on the assumptions made pertaining to lifetime, labor hour cost, overhead, equipment costs, repair costs, discounts for ordering in large quantities, and electricity efficiency. The assumptions in this economic analysis include:

replacing HPS luminaires after 3 years, LED luminaires remain in operation for 12 years, labor cost for relamping or retrofitting luminaires is $60, and the costs for replacing high-pressure sodium lamps for 150, 250, and 400 W lamps are $100, $130, and $160, respectively [15].

The economic analysis assumes high-pressure sodium luminaires are replaced every 3 years. This assumption can easily change to reflect a transportation agency's views of scheduling HPS replacements. The assumption of 3 years accounts for the reduction in luminaire lifetime due to vibration and shock, which is prevalent along bridges and overpasses, and spot replacement of HPS luminaires. In contrast, some transportation agencies wait until the HPS lamp fails catastrophically, which maximizes the lifetime of each luminaire.

Another key assumption is LED luminaires will remain in operation for a 12-year life expectancy. Many manufacturers claim the life of their luminaire will operate beyond 50,000 hours (approximately 12 years with an annual usage of approximately 4000 hours), however, the most common claim is a 12-year lifetime, and 12 years is a conservative lifetime overall for LED luminaires. Therefore, 12 years was used for the LED luminaire lifetime for the economic analysis.

Perspective on labor costs significantly affects the outcome of the economic analysis. Organizations that do not consider maintenance savings as a large factor to their organization will not likely find LED luminaires beneficial. For example, City Utilities in Springfield, MO, replaces traditional street lighting technology on the downtime of their line workers. City Utility policy states that there must be line workers on duty 24 hours per day, 7 days per week in order to respond to outages and emergencies. Therefore, when City Utilities economically analyzed LED luminaires, the results did not favor LED luminaires because the avoided maintenance costs were not included in economic analysis. It is essential for each agency to consider their perspective on replacing or repairing luminaires when performing an economic analysis.

Labor cost to retrofit or relamp a light pole with an LED or an HPS luminaire was assumed to be $60 per luminaire. With lighting labor costs around $25–$35 per hour, the labor cost was averaged and doubled to $60 in order to account for overhead, equipment cost, setup, and travel time to estimate a conservative labor cost.

The costs for replacing high-pressure sodium luminaires vary by the wattage of the lamp being replaced. For the lowest wattage bulb, a $100 cost is used which is based on related LED luminaire analyses. The costs of 250 and 400 W bulbs were estimated to be $130 and $160, respectively. The costs are based on the cost of the lamp being replaced, the cost of labor repairing the lamp's ballast, and the cost of vehicles and equipment to travel to and reach the luminaire.

As previously mentioned, costs may be reduced once roadway lighting demand shifts its focus solely toward LEDs. Economies of scale will then be realized, which similarly occurred in LED traffic signal indicators, and prices of LED luminaires will decrease significantly.

3.1. Life cycle analysis

To determine economic feasibility of LEDs, all costs to install, operate, and dispose of the luminaire are included in the analysis. The installation and disposal costs are accounted for in the retrofitting and relamping labor cost. In addition, the cost of powering the luminaire was calculated based on a sample of actual energy consumption. The actual energy consumption was then extrapolated to other luminaires based on relative wattages between the luminaires which energy consumption was known and other luminaires. Energy consumption for HPS luminaires was calculated using system wattages.

In order to make a fair comparison between HPS luminaires with assumed lifetimes of 3 years and LED luminaires with expected lifetimes of 12 years, the total cost to install and operate a luminaire was annualized. This allows for a fair economic comparison between products with varying lifetimes. An expected project return of 3% was used to annualize costs.

Using information from **Table 2–4**, the annualized costs of LED luminaires is equivalent to or approaching equivalency to HPS lamps. This evaluation of the luminaires was based on pricing for small purchase orders, except for the manufacturer of LED E, which quoted a discounted price for orders of 1000 or more luminaires.

3.2. Replacement period analysis

A potential methodology to level the roadway lighting expenditures while transitioning from HPS luminaires to LED luminaires would be to slowly phase in LED luminaires. By transitioning to LEDs at a rate of the inverse of the expected lifetime of LED luminaires, the annual investment in LEDs is uniform. For example, if LEDs are rated to last for 12 years of use, then 1/12 of lamps should be replaced with LEDs every year. This allows for approximately constant replacement of LED luminaires once the transition from HPS is completed because the failure rate of the LED luminaires will be evenly distributed throughout 12 years.

It is recommended to replace the LED luminaires in large, continuous sections. This will allow for more consistency in overhead street lighting for long sections of road. This will prevent the need to change between the high-pressure sodium and LED luminaires.

Life cycle analysis (150 W equivalents)						
Luminaire	**150W HPS**	**LED A**	**LED B**	**LED C**	**LED D**	**LED E**
Price	$100.00	$695.00	$695.00	$732.00	$700.00	$592.00
Expected lifetime (years)	3	12	12	12	12	12
Project rate of return	3%	3%	3%	3%	3%	3%
Pole installation costs	0	0	0	0	0	0
Relamping/retrofit labor costs	$60.00	$60.00	$60.00	$60.00	$60.00	$60.00
Initial cost per life cycle	$160.00	$755.00	$755.00	$792.00	$760.00	$652.00
Annual electricity consumption	$29.28	$26.91	$28.69	$29.36	$25.80	$32.03
Annualized cost	**$85.84**	**$102.76**	**$104.54**	**$108.92**	**$102.15**	**$97.53**

Table 2. Economic analysis of 150 W equivalent luminaires [15].

Life cycle analysis (250 W equivalents)			
Luminaire	**250W HPS**	**LEF F**	**LED G**
Price	$130.00	$700.00	$712.00
Expected lifetime (years)	3	12	12
Project rate of return	3%	3%	3%
Pole installation costs	0	0	0
Relamping/retrofit labor costs	$60.00	$60.00	$60.00
Initial cost per life cycle	$190.00	$760.00	$772.00
Annual electricity consumption	$48.80	$40.26	$44.48
Annualized cost	**$115.97**	**$116.61**	**$122.04**

Table 3. Economic analysis of 250 W equivalent luminaires [15].

Life cycle analysis (400 W equivalents)		
Luminaire	**400W HPS**	**LED H**
Price	$160.00	$800.00
Expected lifetime (years)	3	12
Project rate of return	3%	3%
Pole installation costs	0	0
Relamping/Retrofit labor costs	$60.00	$60.00
Initial cost per life cycle	$220.00	$860.00
Annual electricity consumption	$78.08	$66.72
Annualized cost	**$155.86**	**$153.12**

Table 4. Economic analysis of 400 W equivalent luminaires [15].

3.3. Sensitivity analysis

Figures 2 and **3** demonstrate the sensitivity of one LED luminaire and one HPS luminaire's annualized cost to changes of four variables: luminaire price, expected luminaire lifetime, relamping, retrofit labor cost, and annual electricity consumption. Each variable varies between 75 and 125% of the original value, in 12.5% intervals. The sensitivity analysis determined the variables with the greatest impact on the annualized cost of LED luminaires. In addition, an incremental economic analysis was performed. The results of the incremental analysis are displayed in **Table 5**. This analysis used the same values as the sensitivity analysis but calculated the change in annual worth per 1% change in each variable. Due to the nonlinearity of the expected lifetime variable, the incremental analysis results of this variable were averaged.

3.4. Sensitivity analysis results

The results of the sensitivity analyses in **Figures 2** and **3** contrast the differences between HPS and LED luminaires as costs change. LED luminaires are significantly less sensitive to

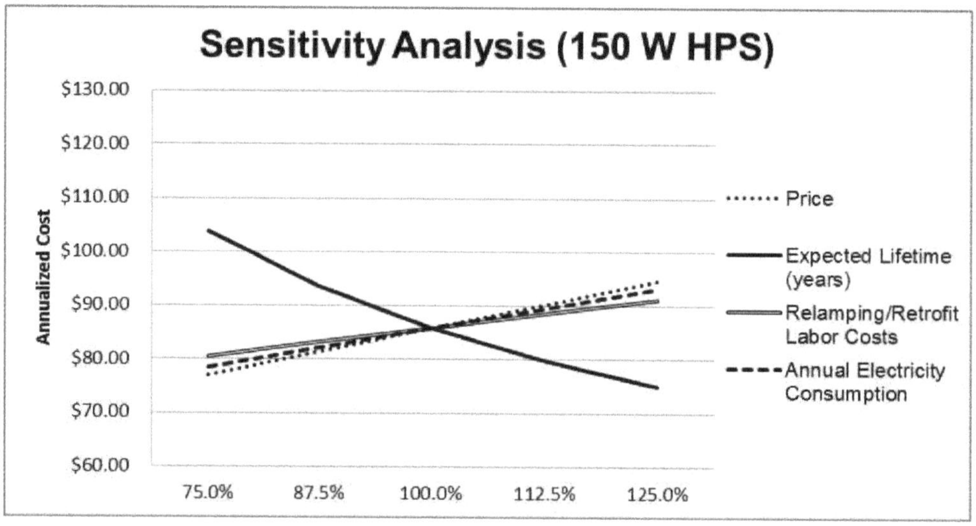

Figure 2. 150 W HPS sensitivity analysis.

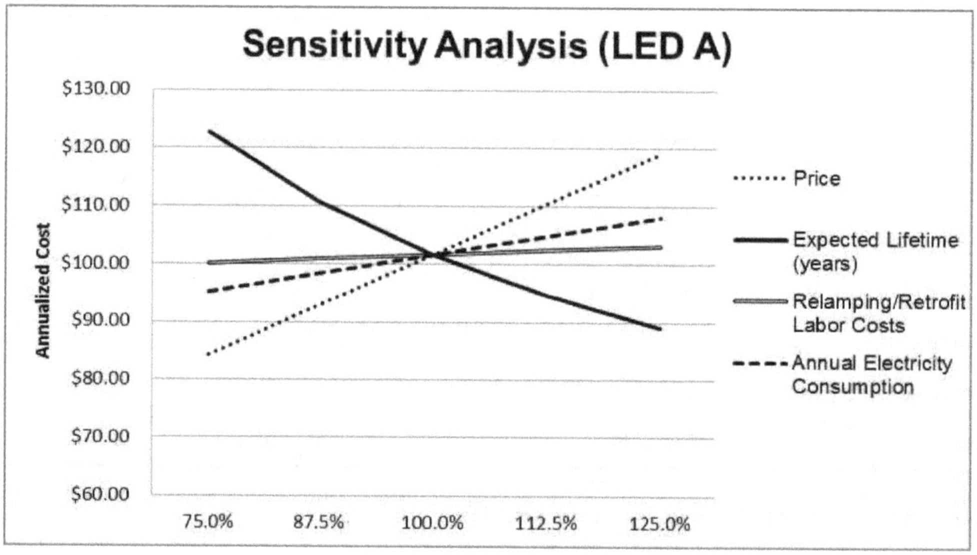

Figure 3. LED a sensitivity analysis.

changes in retrofitting costs, which consist mostly of labor costs. However, LED luminaires are significantly more sensitive to changes in the expected lifetime of the luminaire. Changes in the price of the luminaires linearly impact the annualized cost of the respective luminaire. Changes in each luminaire's expected lifetime result in an inverse exponential change in the annualized cost of the luminaire. Thus, the greater the deviation of the actual lifetime from the expected lifetime, the exponentially greater impact the life of the luminaire has on the annualized cost of the luminaire. Therefore, it is imperative for estimates of an LED luminaire's expected lifetime to be accurate.

The results of the economic sensitivity analysis show the change in annualized cost per 1% change in a variable value. For example, if the price of LED A decreased by 10%, the annualized cost decreases by $7.00. Due to the nonlinearity of the expected lifetime variable, the incremental sensitivity analysis was linearly approximated in order to compare results across

Incremental economic sensitivity analysis				
Luminaire	**Price**	**Expected lifetime (years)**	**Relamping/Retrofit labor costs**	**Annual electricity consumption**
150W HPS	$0.35	($0.56)	$0.21	$0.29
LED A	$0.70	($0.66)	$0.06	$0.27
LED B	$0.70	($0.66)	$0.06	$0.29
LED C	$0.74	($0.69)	$0.06	$0.29
LED D	$0.70	($0.66)	$0.06	$0.26
LED E	$0.59	($0.57)	$0.06	$0.32
250W HPS	$0.46	($0.67)	$0.21	$0.49
LED F	$0.70	($0.66)	$0.06	$0.40
LED G	$0.72	($0.67)	$0.06	$0.44
400W HPS	$0.57	($0.77)	$0.21	$0.78
LED H	$0.80	($0.75)	$0.06	$0.67

Table 5. Incremental sensitivity analysis of HPS and LED luminaires.

all variables. The results of the incremental analysis provide a starting point for effective estimation of annualized costs to account for changes in variable values.

3.5. Energy consumption and environmental impact analysis

Energy consumption data was obtained on a studied luminaire (LED A) at two separate intersections. Both intersections were located in St. Louis, MO. Energy consumption data was normalized to account for days in each month, hours of operation in each month, and the number of luminaires operated at each intersection. Energy consumption data was separated by month and analyzed. **Figure 4** depicts the energy consumption in Watts per luminaire per month.

Figure 4 shows the increase in electricity consumption between October and December, which endures through the month of February. The increase in consumption at this time period averages to 32%. This increase is independent of the duration which the lights operate. The approved product list process section suggests studying this effect further on more luminaires by assessing each luminaire during both summer and winter seasons.

The sharp decrease in March in consumption at the intersection of Route 30 and Main Drive is due to a traffic crash that removed the pole for a period of time. With no replacement LED in stock, one had to be ordered.

Energy consumption was also measured to determine the energy savings of LED luminaires. Our analysis shows an actual energy savings of 11%, which is for 150 W equivalent luminaires. Information was unable to be obtained for equivalent LED power consumption data for 250 W or 400 watt HPS luminaires.

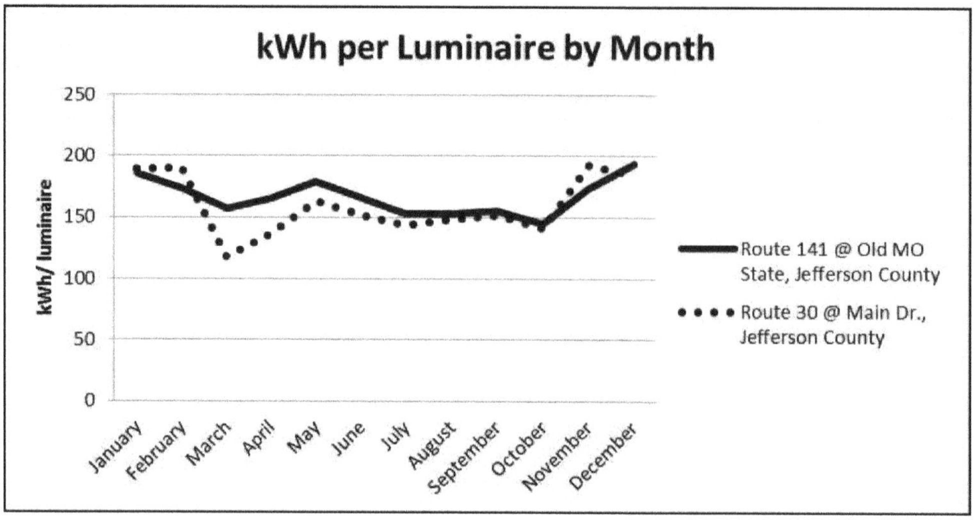

Figure 4. Electricity consumption per luminaire by month.

For a 150 W HPS lamp, with a system rating of 183 watts, the equivalent energy savings is 80.5 kWh per year. According to an EPA study from 2000, the average electrical generation portfolio releases 1.341 lbs (0.608 kg) of CO_2 into the atmosphere per kWh of electricity consumed [16]. Therefore, replacing one 150 Watt HPS lamp with the LED A luminaire avoids the release of approximately 108 lbs of CO_2 into the atmosphere.

4. Conclusions and future work

Performance and cost are major issues when considering a change in technologies such as the transition to the use of LED roadway luminaires. Performance was a major issue in early development of LED roadway luminaires. Most manufacturers invested in product development to ensure that LED roadway luminaires performed at similar or higher performance levels as the HPS roadway luminaires. These initial investments were focused at 30-foot mounting height luminaires and have in the recent past moved toward mounting heights of 40 feet (12.2 m) or higher.

Performance of the LED roadway luminaire, when compared to the current preferred HPS roadway luminaire, has seen improvements over the past few years. Impacted parties (like manufacturers, public agencies, utilities, etc.) have joined together with the intent of producing an equivalent LED roadway luminaire that can be used. Manufacturers have invested in producing new generations of LED roadway luminaires that continue to close the gap between the HPS and LED roadway luminaire. Local agencies and utilities continue to evaluate and report findings on these new generations. Their performance improvements have led some agencies like the City of Los Angeles to make major investments in the transition to LED roadway luminaires.

Results from this research reveal that LED luminaires are less sensitive to changes in retrofitting costs (consisting mostly of labor costs). However, LED luminaires are more sensitive to changes in the expected lifetime of the luminaire. Changes in the price of the luminaires linearly impact the annualized cost of the respective luminaire and changes in each luminaire's expected lifetime result in an inverse exponential change in the annualized cost of the luminaire. Based on these findings, it is essential that life cycle costs for the lifetime of LED luminaires be as accurate as possible. Moreover, the economic sensitivity analysis reveals that incremental analysis provides an effective mechanism for estimating annualized cost.

Energy and environmental analysis shows promising results as well. For a 150-W HPS lamp, with a system rating of 183 W, the equivalent energy savings is 80.5 kWh per year. According to an EPA study from 2000, the average electrical generation portfolio releases 1.341 lbs (0.608 kg) of CO_2 into the atmosphere per kWh of electricity consumed. Therefore, replacing one 150 W HPS lamp with the LED A luminaire avoids the release of approximately 108 lbs of CO_2 into the atmosphere.

Based on our analysis, LED luminaires are a promising technology for replacement of high-pressure sodium lamps. As the technology matures, more robust analysis will confirm the efficacy of the approach.

Acknowledgements

This project was partially funded through the Missouri Department of Transportation (TRyy1101) and the data used in this case was originally published in the corresponding final report. The authors would like to thank Tom Ryan, Dr. A. Curt Elmore, and Dr. Ruwen Qin for their input and guidance throughout the research project. We also want to thank Julie Stotlemeyer and Jen Harper, MoDOT, for their valuable assistance.

Author details

Sean Schmidt and Suzanna Long*

*Address all correspondence to: longsuz@mst.edu

Department of Engineering Management and Systems Engineering, Missouri University of Science and Technology, Rolla, MO, USA

References

[1] *LED Street Lighting Efficiency Program*. Los Angeles: City of Los Angeles – Bureau of Street Lighting, 2011. http://bsl.lacity.org. Accessed December 2011.

[2] Pacific Northwest National Laboratory. *Demonstration Assessment of Light-Emitting Diode (LED) Roadway Lighting*. Washington D.C.: U.S. Department of Energy, 2009.

[3] Radetsky LC. *Specifier Reports: Streetlights for Collector Roads*. Troy, Rensselaer Polytechnic Institute NY, 2010.

[4] Radetsky LC. *Specifier Reports: Streetlights for Local Roads*. Troy, Rensselaer Polytechnic Institute NY, 2011.

[5] Illumination Engineering Society. Roadway Lighting. Tech. No. RP-8-OOR2005. NY, 2000.

[6] Avrenli KA, Benekohal RF, Medina JC. *LED Roadway Lighting*. Urbana: University of Illinois at Urbana-Champaign, Department of Civil and Environmental Engineering, 2012.

[7] Pacific Northwest National Laboratory. *Demonstration Assessment of Light-Emitting Diode (LED) Parking Lot Lighting*. Washington D.C., U.S. Department of Energy 2011.

[8] Beckwith D, Zhang X, Smalley E, Chan L, Yand M. LED streetlight application assessment project. Transportation Research Record, 2011, pp. 65–75.

[9] Ellis RD, Amos SJ. Development of Work Zone Lighting standards for nighttime highway work. Transportation Research Record. 1996, pp. 35–42.

[10] Kostic M, Djokic L. Recommendations for energy efficient and visually acceptable street lighting. Energy. 2009; 34(10): pp. 1565–72.

[11] Kinzey BR, Royer MP, Hadjian M, Kauffman R. *Demonstration of LED Street Lighting in Kansas City, MO*. Washington D.C.; U.S. Department of Energy 2013.

[12] Pacific Northwest National Laboratory. *LED Street Lighting*. Washington D.C.; U.S. Department of Energy 2008.

[13] Gil-de-Castro A, Moreno-Munoz A, Larsson A, JJG de la Rosa, Bollen. LED street lighting: A power quality comparison among street light technologies. Lighting Research and Technology. 2013; 45: pp. 710–728.

[14] Pizutti S, Annunziato M, Moretti F. Smart street lighting management. Energy Efficiency. 2013; 6: pp. 609–616.

[15] Long S, Qin R, Elmore C, Ryan T, Schmidt S. *LED Roadway Luminaires Evaluation (TRyy1101)*. Research Report. Missouri Department of Transportation, 2011. https://library.modot.mo.gov/rdt/reports/tryy1101/cmr12011tryy1101.pdf

[16] U.S. Environmental Protection Agency and Department of Energy. *Carbon Dioxide Emissions from the Generation of Electric Power in the United States*. Washington D.C.; U.S. Environmental Protection Agency/U.S. Department of Energy 2000.

6

Information-Theoretic Clustering and Algorithms

Toshio Uchiyama

Abstract

Clustering is the task of partitioning objects into clusters on the basis of certain criteria so that objects in the same cluster are similar. Many clustering methods have been proposed in a number of decades. Since clustering results depend on criteria and algorithms, appropriate selection of them is an essential problem. Recently, large sets of users' behavior logs and text documents are common. These are often presented as high-dimensional and sparse vectors. This chapter introduces information-theoretic clustering (ITC), which is appropriate and useful to analyze such a high-dimensional data, from both theoretical and experimental side. Theoretically, the criterion, generative models, and novel algorithms are shown. Experimentally, it shows the effectiveness and usefulness of ITC for text analysis as an important example.

Keywords: information-theoretic clustering, competitive learning, Kullback-Leibler divergence, Jensen-Shannon divergence, clustering algorithm, text analysis

1. Introduction

Clustering is the task of partitioning objects into clusters on the basis of certain criteria so that objects in the same cluster are similar. It is a fundamental procedure to analyze data [1, 2].

Clustering is *unsupervised* and different from *supervised* classification. In supervised classification, we have a set of *labeled* data (belong to predefined classes), train a classifier using the labeled data (*training set*), and judge which class a new object belongs to by the classifier. In the case of clustering, we find meaningful clusters without using any labeled data and group a given collection of unlabeled data into them. Clustering can also help us to find meaningful classes (labels) for supervised classification. Since it is more difficult to prepare the training set for larger data sets, recently unsupervised analysis of data such as clustering becomes more important.

For example, **Table 1** *user-item matrix* shows which item a user bought. When considering the data as a set of feature vectors for users, we can find a lot of types of users' behavior by

	item1	item2	item3	item4	item5
User1	3	0	0	5	0
User2	0	1	2	0	0
User3	1	0	0	0	2
User4	0	0	1	0	0

$$\Rightarrow \begin{pmatrix} 3 & 0 & 0 & 5 & 0 \\ 0 & 1 & 2 & 0 & 0 \\ 1 & 0 & 0 & 0 & 2 \\ 0 & 0 & 1 & 0 & 0 \end{pmatrix}$$

Table 1. Consumption behavior of users.

	Document1	Document2	Document3	Document4	Document5
Word1	3	0	0	5	0
Word2	0	1	2	0	0
Word3	1	0	0	0	2
Word4	0	0	1	0	0

Table 2. Word frequencies in documents (bag-of-words feature representation).

clustering. It is also possible to analyze data as a set of feature vectors for items. From *word-document matrix* in **Table 2**, both document clusters and word clusters could be extracted.

Many clustering methods have been proposed in a number of decades. Those include k-means algorithm [3], competitive learning [4], spherical clustering [5], spectral clustering [6], and maximum margin clustering [7]. Since clustering results depend on criteria and algorithms, appropriate selection of them is an essential problem. Large sets of users' behavior logs and text documents (as shown in **Tables 1** and **2**) are common recently. These are often presented as high-dimensional and sparse vectors. This chapter introduces information-theoretic clustering [8] and algorithms that are appropriate and useful to analyze such a high-dimensional data.

Information-theoretic clustering (ITC) uses Kullback-Leibler divergence and Jensen-Shannon divergence to determine its criterion, while k-means algorithm uses the sum of squared error as criterion. This chapter explains ITC by contrasting these two clustering techniques (criteria and algorithms), because there are a number of interesting similarities between them. There exists difficulty in algorithms for ITC. We explain the details of it and propose novel algorithms to overcome.

Experimental results for text data sets are presented to show the effectiveness and usefulness of ITC and novel algorithms for it. In experiments, maximum margin clustering and spherical clustering are used to compare. We also provide the evidence to support the effectiveness of ITC by detailed analysis of clustering results.

2. The sum-of-squared-error criterion and algorithms

Given a set of M-dimensional input vectors $\mathcal{X} = \{x^i | x^i \in R^M, i = 1, \ldots, N\}$ where N is the number of vectors, clustering is the task of assigning each input vector x^i a cluster label $k(k = 1, \ldots, K)$ to

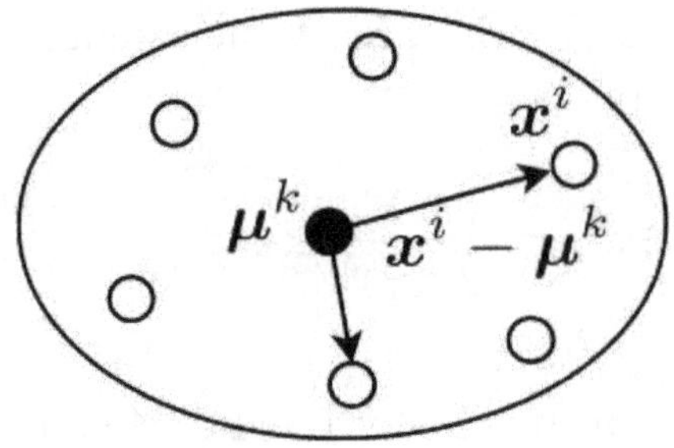

Figure 1. Input vectors and the mean vector in C^k.

partition them into K clusters $\mathcal{C} = \{C^1,\ldots,C^K\}$. The *sum-of-squared-error criterion* [9] is a simple and widely used criterion for clustering.

Let $\boldsymbol{\mu}^k$ be the mean of the input vectors $\boldsymbol{x}^i$ which belong to the cluster C^k (see **Figure 1**). Then, the error in C^k is the sum of squared lengths of the differential (= "error") vectors $\|\boldsymbol{x}^i-\boldsymbol{\mu}^k\|^2$ and the sum-of-squared-error criterion about all clusters (*within-cluster* sum of squares) is defined by

$$J_W = \sum_{k=1}^{K} \sum_{\boldsymbol{x}^i \in C^k} \|\boldsymbol{x}^i-\boldsymbol{\mu}^k\|^2. \tag{1}$$

J_W is the objective function (criterion) to be minimized in clustering based on this criterion.

Also, we define the sum of squares of *between-cluster* J_B and *total* J_T as

$$J_B = \sum_{k=1}^{K} N_k\|\boldsymbol{\mu}^k-\boldsymbol{\mu}\|^2, \; J_T = \sum_{i=1}^{N} \|\boldsymbol{x}^i-\boldsymbol{\mu}\|^2, \tag{2}$$

respectively, where N_k is the number of input vectors x^i in C^k (i.e., $N = \sum_{k=1}^{K} N_k$) and

$$\boldsymbol{\mu}^k = \frac{1}{N_k} \sum_{\boldsymbol{x}^i \in C^k} \boldsymbol{x}^i, \; \boldsymbol{\mu} = \frac{1}{N}\sum_{i=1}^{N} \boldsymbol{x}^i. \tag{3}$$

It follows from these definitions that the total sum of squares is the sum of the within-cluster sum of squares and the between-cluster sum of squares:

$$J_T = J_W + J_B. \tag{4}$$

Since the mean of the all input vectors $\boldsymbol{\mu}$ is derived from $\mathcal{X} = \{\boldsymbol{x}^1,\ldots,\boldsymbol{x}^N\}$ [see Eq. (3)], J_T does not depend on clusters $\mathcal{C}$ [see Eq. (2)] and is constant for the given input vectors $\mathcal{X}$. Therefore, minimization of J_W is equivalent to maximization of J_B. In this sense, clustering based on minimizing this criterion J_W works to find separable clusters each other.

2.1. Generative model

In the background of the clustering based on the objective function (criterion) J_W, there exists assumption of Gaussian distribution about input vectors [10].

Suppose that there are clusters $C^k(k = 1,\ldots,K)$, which generates input vectors by the conditional probability density function:

$$p(\boldsymbol{x}^i|\boldsymbol{x}^i \in C^k) = \frac{1}{(2\pi\sigma_k^2)^{M/2}} \exp\left(-\frac{\|\boldsymbol{x}^i - \boldsymbol{\mu}^k\|^2}{2\sigma_k^2}\right), \tag{5}$$

where σ_k is a standard deviation of the cluster C^k and M is the number of dimension of $\boldsymbol{x}^i$. In followings, we assume that σ_k is constant value σ for all clusters $C^k(k = 1,\ldots,K)$. Considering independence of each generation, joint probability density function for the input vectors $\mathcal{X}$ becomes

$$p(\mathcal{X}|\mathcal{C}) = \prod_{k=1}^{K} \prod_{\boldsymbol{x}^i \in C^k} \frac{1}{(2\pi\sigma^2)^{M/2}} \exp\left(-\frac{\|\boldsymbol{x}^i - \boldsymbol{\mu}^k\|^2}{2\sigma^2}\right), \tag{6}$$

where $\mathcal{C}$ indicate cluster information that specifies which input vector $\boldsymbol{x}^i$ belongs to cluster C^k. Taking the logarithm of Eq. (6) yields

$$\text{ln}p(\mathcal{X}|\mathcal{C}) = -\frac{NM}{2}\log(2\pi\sigma^2) - \frac{1}{2\sigma^2}\sum_{k=1}^{K}\sum_{\boldsymbol{x}^i \in C^k}\|\boldsymbol{x}^i - \boldsymbol{\mu}^k\|^2. \tag{7}$$

Since σ is constant, the maximization of Eq. (7) is equivalent to the minimization of

$$\sum_{k=1}^{K}\sum_{\boldsymbol{x}^i \in C^k}\|\boldsymbol{x}^i - \boldsymbol{\mu}^k\|^2. \tag{8}$$

which is nothing more or less than the objective function (criterion) J_W. Therefore, under the assumption of Gaussian distribution about input vectors, clustering based on Eq. (8) works to find the most probable solution $\mathcal{C}$.

2.2. Algorithms

2.2.1. k-means algorithm

k-means [3, 11] is well-known algorithm for clustering based on the sum-of-squared-error criterion. Main idea of this algorithm is as follows. In the objective function J_W (1), error for vector $\boldsymbol{x}$ is calculated by $\|\boldsymbol{x} - \boldsymbol{\mu}^k\|^2$ where $\boldsymbol{\mu}^k$ is the mean of cluster C^k to which $\boldsymbol{x}$ belongs. If $\|\boldsymbol{x} - \boldsymbol{\mu}^t\|^2 < \|\boldsymbol{x} - \boldsymbol{\mu}^k\|^2$, changing the cluster from C^k to C^t can reduce the objective function J_W.

We introduce **weight vector** $\boldsymbol{w}^k(k = 1,\ldots,K)$ $(\mathcal{W})$ that represent cluster C^k to implement the idea mentioned above. The weight vector $\boldsymbol{w}^k$ involves mean vector $\boldsymbol{\mu}^k$ and prototype vector of cluster C^k. As illustrated in **Figure 2**, the idea of *k-means* is alternative repetition of two steps "(a) Update weights" (calculating mean $\boldsymbol{\mu}^k$ as weight vector $\boldsymbol{w}^k$) and "(b) Update clusters" (allocating input vector $\boldsymbol{x}^i$ to a cluster C^k on the basis of minimum length from weight vectors $\boldsymbol{w}^k$). Note that **Figure 2b** is a Voronoi tessellation determined by weight vectors $\boldsymbol{w}^k$, which are usually called *prototype vector* in this context.

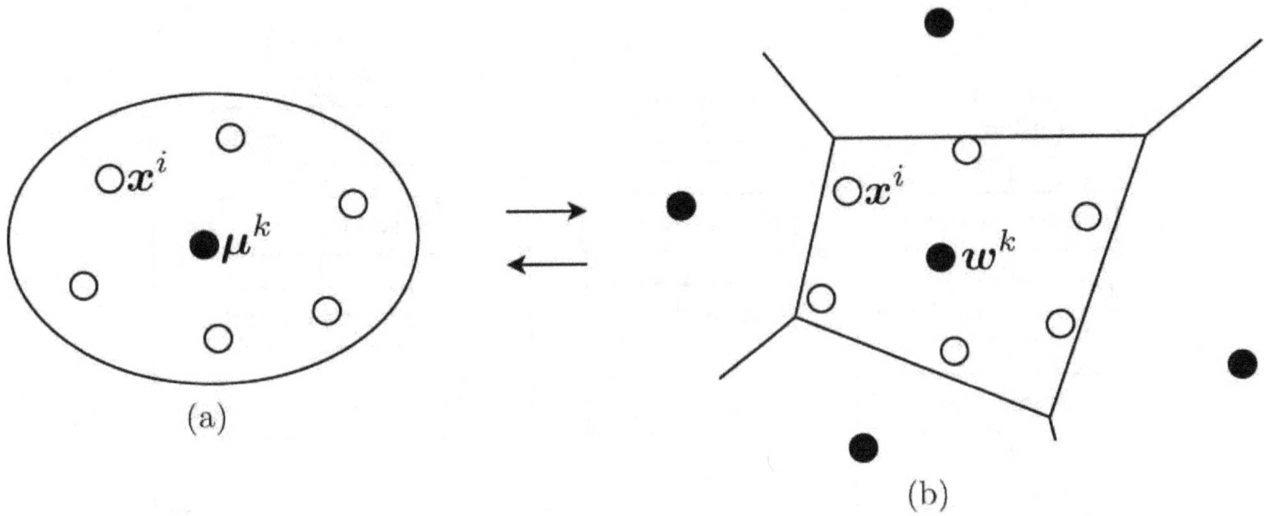

Figure 2. Two steps in k-means algorithm. (a) Update weights. (b) Update clusters.

Figure 3a is a flow chart of k-means algorithm to which processes of initialization and termination are added. As a matter of fact, clustering is closely related to vector quantization. Vector quantization means mapping input vectors to a codebook that is a set of weight vectors (prototype vectors). When using quantization error E_Q:

$$E_Q = \sum_{i=1}^{N} \min_k ||x^i - w^k||^2, \tag{9}$$

clusters $\mathcal{C}$ determined by a local optimal solution of vector quantization $\mathcal{W}$ is a local optimal solution of clustering problem [12]. In this sense, clustering can be replaced by vector quantization and vice versa. We can write a flow chart for vector quantization as **Figure 3b**, but we also find this chart (b) as k-means algorithm. Furthermore, *LBG algorithm* [13], which is well known for vector quantization, is based on an approach of Lloyd [3] (one of original papers for k-means algorithm). These facts show a close relationship between clustering and vector quantization.

Initialization is important, because k-means algorithm converges to a local optimal solution which depends on an initial condition (a set of weights or clusters). If we initialize weights $\mathcal{W}$ by randomly selecting them from input vectors, it may converge to a very bad local optimal solution with high probability. *Random labeling* that randomly assigns cluster labels $\mathcal{C}$ to input vectors may lead to better solutions than random selection of weights. The initialization *Random labeling* can also be used for charts (b) and (c) in **Figure 3** by replacing "Initialize weights" step to "Initialize clusters" and "Update weights" steps. For directly initializing weights, splitting algorithm [13] and *k-means ++* [14] were known.

2.2.2. Competitive learning

Competitive learning [4, 11] is a learning method for vector quantization and also utilized for clustering. While k-means algorithm updates all weights $\mathcal{W}$ by batch processing, competitive learning updates one weight w at a time to reduce a part of the quantization error Q_E (see **Figure 3c**) as

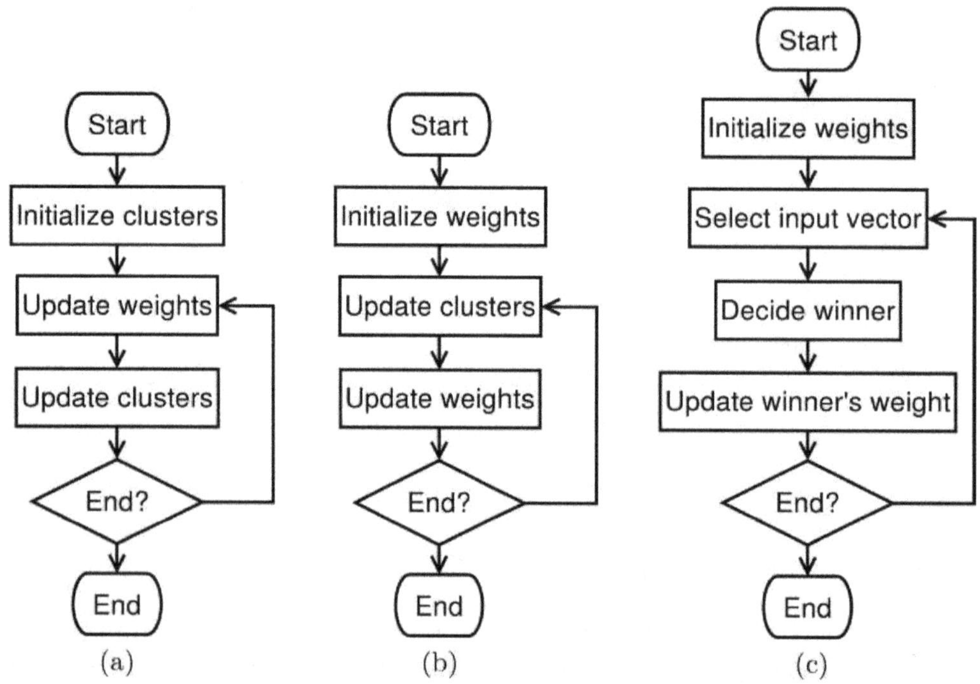

Figure 3. Flow charts of algorithms based on sum-of-squared-error criterion. (a) k-means1, (b) k-means2, and (c) competitive learning.

1. Select one input vector $\boldsymbol{x}$ randomly from $\mathcal{X}$.
2. Decide a winner $\boldsymbol{w}^c$ from $\mathcal{W}$ by

$$c = \arg\min_k \|\boldsymbol{x} - \boldsymbol{w}^k\|^2 \text{ (If there are several candidates, choose the smallest k).} \tag{10}$$

3. Update the winner's weight $\boldsymbol{w}^c$ as

$$\boldsymbol{w}^c \leftarrow (1-\gamma)\boldsymbol{w}^c + \gamma \boldsymbol{x}, \tag{11}$$

where γ is a given learning rate (e.g., 0.01–0.1).

Though the *winner-take-all* update in Step 3 (**Figure 4**) that reduces partial error $\|\boldsymbol{x}-\boldsymbol{w}^c\|^2$ in steepest direction does not always reduce the total quantization error E_Q, repetition of the update can reduce E_Q on the basis of **stochastic gradient decent** method [15, 16]. For termination condition, maximum number of times of iteration N_r (*the number of maximum repetitions*) can be used. After termination, the step of deciding clusters $\mathcal{C}$ like **Figure 2b** is required for clustering purpose.

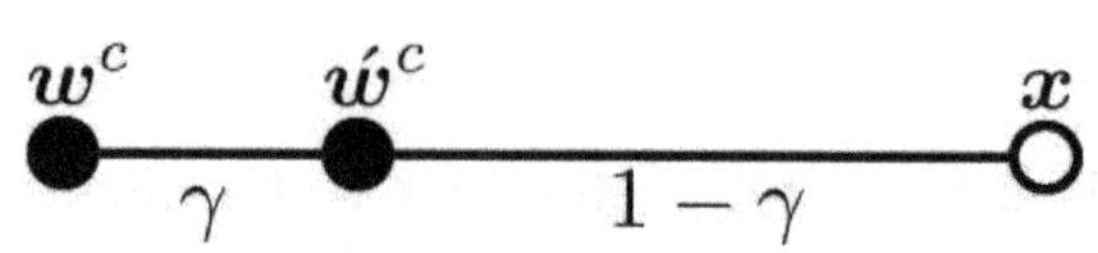

Figure 4. Update of the winner's weight in competitive learning.

Against natural expectations, competitive learning outperforms k-means without any contrivance in most cases. Furthermore, information obtained in learning process allows us to improve its performance. Splitting rule [12] utilizes the number of each weight w wins to estimate density around it. As **Figure 5a, b** shows, higher density of input vectors around makes the weight vector w^a win more frequently than w^b.

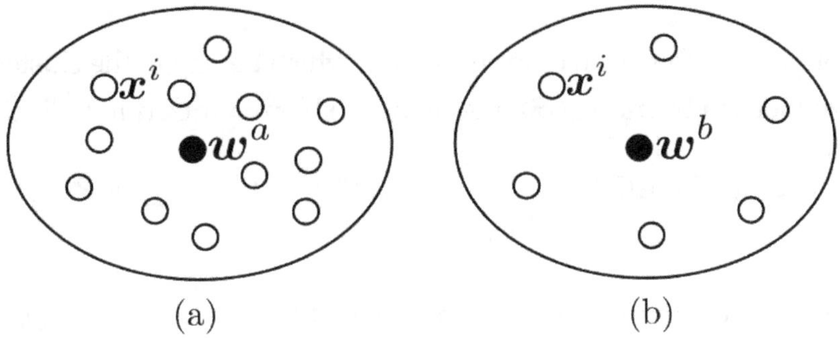

Figure 5. Density of input vectors around a weight vector.

Splitting rule in competitive learning [12] aims to overcome the problem of discrepancy between distribution of input vectors $\mathcal{X}$ and that of weight vectors $\mathcal{W}$. The discrepancy causes a few weight vectors w monopolize $\mathcal{X}$ and leads to a solution of very poor quality, but it is impossible to figure out the distribution of input vectors beforehand. Accordingly, this splitting rule distributes weight vectors w in learning process as

1. One weight vector w^1 with a variable τ^1 is set. τ^1 denotes how many times weight vector w^1 wins and is initialized to 0.
2. Select one input vector x, decide winner w^c, and update the winner's weight w^c.
3. Add 1 to τ^c. If $\tau^c = \theta$ and the current number of weights K' is less than K, generate a new weight vector w which is the same as the winner w^c and clear τ of both to 0, where θ is the threshold of times for splitting.
4. Repeat 2 and 3 until termination condition is true.

3. Information-theoretic clustering and algorithms

Information-theoretic clustering (ITC) [8] is closely related to works about **distributional clustering** [17–19] and uses Kullback-Leibler divergence and Jensen-Shannon divergence to determine its criterion. Though there exists difficulty in algorithms and effectiveness for high-dimensional count data (e.g., text data), its definition and properties are similar to those of the sum-of-squared-error criterion. The main contributions of this chapter are to present the technique to overcome the difficulty and effectiveness of ITC.

Let $\mathcal{X} = \{x^i | x^i \in R_+^M, i = 1, \ldots, N\}$ be a set of M-dimensional input vectors (N denote the number of input vectors), where elements of vectors x are nonnegative real numbers. We define a l^1-norm of input vector $t^i(=\sum_m|x^i_m|)$, normalized input vectors $\boldsymbol{p}^i = \boldsymbol{x}^i/t^i$, and an input probability distribution P^i whose mth random variable takes the mth element of $\boldsymbol{p}^i(=p^i_m)$. Let $\mathcal{P} = \{P^1, \ldots, P^N\}$ be a set of input distributions (input data).

Suppose that we assign each distribution P^i a cluster label $k(k = 1, \ldots, K)$ to partition them into K clusters $\mathcal{C} = \{C^1, \ldots, C^K\}$.

Let $\overline{P}^k$ be the distributions on the mean of input data P^i which belong to the cluster C^k (see **Figure 6**). Then, the generalized Jensen-Shannon (JS) divergence to be minimized in C^k is defined by

$$D_{\mathrm{JS}}(\{P^i|P^i \in C^k\}) = \sum_{P^i \in C^k} \pi^i D_{\mathrm{KL}}(P^i \| \overline{P}^k), \ \overline{P}^k = \sum_{P^i \in C^k} \pi^i P^i, \tag{12}$$

where N_k is the number of distributions P^i in cluster C^k (i.e., $N = \sum_{k=1}^K N_k$), $D_{\mathrm{KL}}(P^i \| \overline{P}^k)$ is the Kullback-Leibler (KL) divergence to the mean distribution $\overline{P}^k$ from P^i, and π^i is the probability of P^i ($\sum_{P^i \in C^k} \pi^i = 1$). Here $P^i = 1/N_k$. Then, we define *within-cluster* JS divergence JS_W which considers all clusters $C^k(k = 1, \ldots, K)$ as

$$JS_W = \sum_{k=1}^{K} \frac{N_k}{N} D_{\mathrm{JS}}(\{P^i|P^i \in C^k\}) \tag{13}$$

$$= \frac{1}{N} \sum_{k=1}^{K} \sum_{P^i \in C^k} D_{\mathrm{KL}}(P^i \| \overline{P}^k) \tag{14}$$

$$= \frac{1}{N} \sum_{k=1}^{K} \sum_{P^i \in C^k} \sum_{m=1}^{M} p^i_m \log \frac{p^i_m}{\overline{p}^k_m} = \frac{1}{N} \sum_{k=1}^{K} \sum_{P^i \in C^k} \sum_{m=1}^{M} (p^i_m \log p^i_m - p^i_m \log \overline{p}^k_m). \tag{15}$$

The *within-cluster* JS divergence JS_W is the objective function (criterion) of **information-theoretic clustering (ITC)** to be minimized [8]. We also define JS divergence of *between-cluster* JS_B and *total* JS_T as

$$JS_B = D_{\mathrm{JS}}(\{\overline{P}^k | k = 1, \ldots, K\}) = \sum_{k=1}^{K} \pi^k D_{\mathrm{KL}}(\overline{P}^k \| \overline{P}), \ \pi^k = N_k/N, \tag{16}$$

$$JS_T = D_{\mathrm{JS}}(\{P^i | i = 1, \ldots, N\}) = \sum_{i=1}^{N} \pi^i D_{\mathrm{KL}}(P^i \| \overline{P}), \ \pi^i = 1/N, \tag{17}$$

where $\overline{P} = \sum_{i=1}^N \pi^i P^i = 1/N \sum_{i=1}^N P^i$ is the distribution on the mean of all input data. It follows from these definitions that the total JS divergence is the sum of the within-cluster JS divergence and the between-cluster JS divergence [8]:

$$JS_T = JS_W + JS_B. \tag{18}$$

Since JS_T are constant for given input distributions $\mathcal{P}$, minimization of JS_W is equivalent to maximization of JS_B. In this sense, clustering based on minimizing this criterion JS_W works to find separable clusters each other.

The definition and properties of ITC as shown so far are similar to those of the sum-of-squared-error criterion. Those will help us to understand ITC.

3.1. Generative model

In the background of information-theoretic clustering (ITC), there also exists the *bag-of-words* assumption [20] that disregards the order of words in a document. (Since ITC is not limited for document clustering, "word" is just an example of feature.) It means that features in data are conditionally independent and identically distributed, where the condition is a given probability distribution for an input vector. Based on this assumption, we describe a generative probabilistic model related to ITC and make clear the relationship between the model and the objective function (criterion) JS_W.

Let an input vector $\boldsymbol{x} = \{x_1,\ldots,x_m,\ldots,x_M\}$ present a set of the number of observations of mth feature. Suppose that there are clusters $C^k(k = 1,\ldots,K)$ which generates t^i features for a data (= input vector) with the probability distribution $\overline{P}^k = \{\overline{p}_1^k,\ldots,\overline{p}_M^k\}$, and conditional probability of a set of the observation about features in an input vector $\boldsymbol{x}^i$ is expressed by multinomial distribution

$$p(\boldsymbol{x}^i|\boldsymbol{x}^i \in C^k) = A^i \prod_{m=1}^{M} (\overline{p}_m^k)^{x_m^i}, \quad A^i = \frac{t^i!}{x_1^i! \cdot x_2^i! \cdots x_M^i!}, \quad t^i = \sum_{m=1}^{M} |x_m^i|, \tag{19}$$

where A^i is the number of combination of the observation. Assuming independence of each generation, joint probability function for the input vectors $\mathcal{X} = \{\boldsymbol{x}^1,\ldots,\boldsymbol{x}^N\}$ becomes

$$p(\mathcal{X}|C) = \prod_{k=1}^{K} \prod_{x^i \in C^k} A^i \prod_{m=1}^{M} (\overline{p}_m^k)^{x_m^i}, \tag{20}$$

where C indicates cluster information that specifies which input vector $\boldsymbol{x}^i$ belongs to cluster C^k. Taking the logarithm of Eq. (20) yields

$$\ln p(\mathcal{X}|C) = \sum_{i=1}^{N} \log A^i + \sum_{k=1}^{K} \sum_{x^i \in C^k} \sum_{m=1}^{M} \boldsymbol{x}_m^i \log \overline{p}_m^k \tag{21}$$

$$= \sum_{i=1}^{N} \log A^i + \sum_{k=1}^{K} \sum_{p^i \in C^k} \sum_{m=1}^{M} t^i \cdot p_m^i \log \overline{p}_m^k. \tag{22}$$

This is *a generative probabilistic model* related to ITC. If we assume that t^i takes constant value t for all input vectors, maximization of the probability (22) as well as minimization of the objective function JS_W (15) come to the minimization of

$$\frac{1}{N} \sum_{k=1}^{K} \sum_{p^i \in C^k} \sum_{m=1}^{M} -p_m^i \log \overline{p}_m^k = \frac{1}{N} \sum_{k=1}^{K} N_k \sum_{m=1}^{M} -\overline{p}_m^k \log \overline{p}_m^k, \tag{23}$$

for given input distribution $\mathcal{P}$. Here, the relationship $\sum_{p^i \in C^k} p_m^i = N_k \overline{p}_m^k$ is used. Since t^i may not be constant value t, the generative model (22) is not an equivalent model of ITC but the *related*

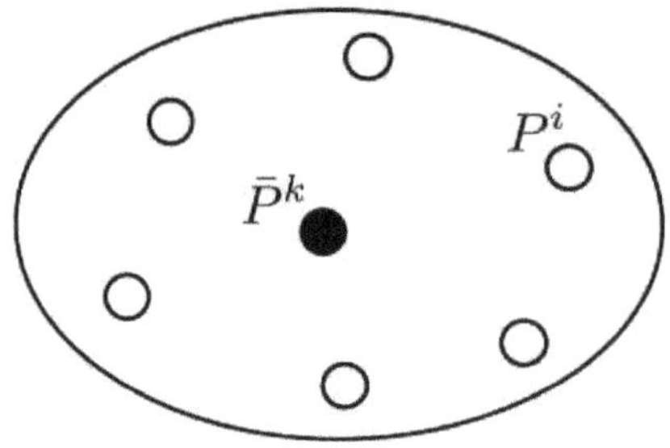

Figure 6. Input distributions and the mean in C^k.

model. This difference comes from the fact that the model treats each observation about features equally, while ITC treats each data (input vector) equally. Though the additional assumption $t^i = t$ is required, ITC works to find the most probable solution $\mathcal{C}$ in the generative probabilistic model. Furthermore, Eq. (23) is also based on the minimization of *entropy* in clusters as Eq. (23) shows. Entropy (specifically, Shannon Entropy) is the expected value of the information contained in each message that is an input distribution here. The smaller entropy becomes, and the more compactly a model can explain observations (input distributions). In this sense, the objective function JS_W (15) presents the goodness of the generative model. The relationship (including difference) between the probabilistic model and the objective function JS_W is meaningful to improve the model and the objective function in future.

Choice of appropriate model for data is important, when analyzing them. For example, large set of text documents contain many kinds of words and are presented as high-dimensional vectors. Taking extreme diversity of documents' topics into account, feature vectors of documents are distributed almost uniformly in the vector space. As known by **"the curse of dimensionality"** [10], most of the volume of a sphere in high-dimensional space is concentrated near the surface, and it becomes not appropriate to choose the model based on Gaussian distribution which concentrates values around the mean. In contrast, ITC on the basis of the multinomial distribution is a reasonable and useful tool to analyze such a high-dimensional count data, because the generative model of ITC is consistent with them.

We introduce **weight distribution** $Q^k(k = 1,\ldots,K)(\mathcal{Q})$ that represent cluster C^k and that involves mean distribution $\overline{P}^k$ and prototype distribution of cluster C^k in a manner similar to that of the sum-of-squared-error (SSE) criterion (see Section2.2.1). **Figure 7** shows relationships between parameters in generative models. Parameters are generated or estimated by other parameters to maximize probability of generative model. For example, clustering is the task to find the most probable clusters $\mathcal{C}$ for given input vectors $\mathcal{X}$ or input distributions $\mathcal{P}$. In **Figure 7b**, constructing a classifier is the task to find $\mathcal{Q}$ for given $\mathcal{P}$ and $\mathcal{C}$ (classes in this context) in training process. Then, it estimates $\mathcal{C}$ for unknown $\mathcal{P}$ using the trained $\mathcal{Q}$. The classifier using multinominal distribution is known as multinominal *Naive Bayes classifier* [21]. As it shows, ITC and Naive Bayes classifier have a close relationship [18].

3.2. Algorithms

There exists difficulty (Appendix A) in algorithms for ITC. We show a novel idea to overcome it.

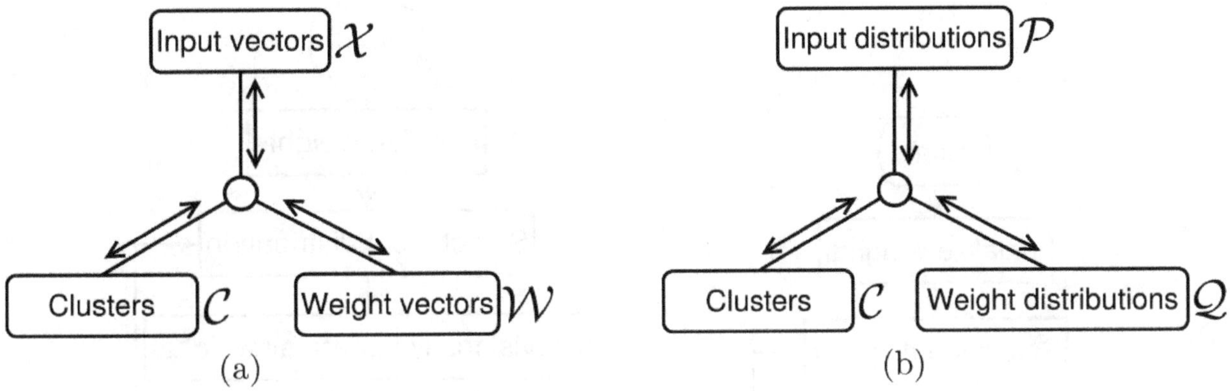

Figure 7. Relationships between model parameters. (a) Clustering based on SSE criterion. (b) Information-theoretic clustering

3.2.1. Competitive learning

When competitive learning decides a winner for an input distribution P, it easily faces the difficulty of calculating KL divergence from P to weight distributions Q (see Appendix A). To overcome this difficulty, we present the idea to change an order of steps in competitive learning (CL). As shown in **Figure 8b**, CL updates all weights (= weight distributions) before deciding winner by

$$Q^k \leftarrow (1-\gamma)Q^k + \gamma P, \tag{24}$$

where γ is a learning rate. Since updated weight distributions $Q^k (k = 1,\ldots,K)$ include all words (features) of input distribution P, it is possible to calculate KL divergence $D_{\mathrm{KL}}(P \| Q^k)$ for all k. In following steps, CL decide a winner Q^c from $\mathcal{Q}$ by

$$c = \arg\min_k D_{\mathrm{KL}}(P \| Q^k) \ \text{(If there are several candidates, choose the smallest k)}, \tag{25}$$

and activate winner's update and discard others. These steps satisfy the CL's requirement that it partially reduces value of objective function JS_W in steepest direction with the given learning rate γ. Here, neither approximation nor distortion is added to the criterion of ITC. Note that updates of weight distributions Q^k before activation are provisional (see **Figure 8b**).

Related work that avoids the difficulty in calculating KL divergence presented **skew divergence** [22]. The skew divergence is defined as

$$s_\alpha(P,Q) = D_{\mathrm{KL}}(P \| \alpha Q + (1-\alpha)P), \tag{26}$$

where $\alpha (0 \le \alpha \le 1)$ is the mixture ratio of distributions. The skew divergence is exactly the KL divergence at $\alpha = 1$. When $\alpha = 1-\gamma$, Eq. (26) becomes similar to Eq. (24). Then, we can rewrite the steps in CL above using the skew divergence as

1. Select one input distribution P randomly from $\mathcal{P}$.
2. Decide a winner Q^c from $\mathcal{Q}$ by

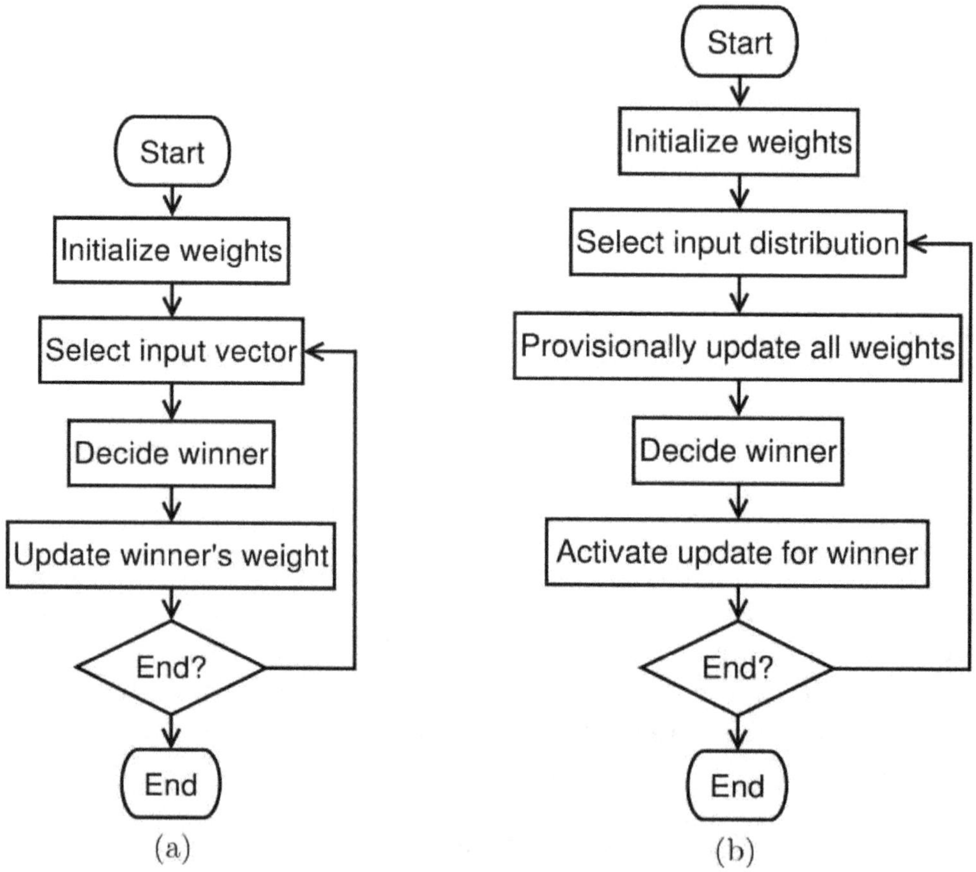

Figure 8. Flow charts of competitive learning. (a) Competitive learning for SSE. (b) Competitive learning for ITC

$$c = \arg\min_k s_\alpha(P, Q^k) \text{ (If there are several candidates, choose the smallest k)}, \tag{27}$$

3. Update the winner's weight distribution Q^c as

$$Q^c \leftarrow (1-\gamma)Q^c + \gamma P, \tag{28}$$

where γ is a learning rate and equal to $1-\alpha$ (α is the mixture ratio for s_α) usually.

Hence, we call this novel algorithm for ITC as **"competitive learning using skew divergence"** (sdCL). In addition, splitting rule in competitive learning [12] can also be applied to this algorithm.

3.2.2. k-means type algorithm

Dhillon et al. [8] proposed *information-theoretic divisive algorithm* which is k-means type algorithm with divisive mechanism and uses KL divergence.[1] However, it still remains the

[1] The algorithm was proposed for feature/word clustering and applied to text classification. Since the algorithm uses document class (labeled data), it cannot be applied to general clustering problem.

difficulty to use KL divergence directly. In such a situation, we propose to use the skew divergence instead of KL divergence in k-means type algorithm as

1. Initialize clusters $\mathcal{C}$ of input distribution $\mathcal{P}$ randomly.
2. Update weight distributions $\mathcal{Q}$ by

$$Q^k = \frac{1}{N_k} \sum_{P^i \in C^k} P^i. \tag{29}$$

3. Update each cluster c of an input distribution P^i by

$$c = \arg\min_k s_\alpha(P^i, Q^k) \text{ (If there are several candidates, choose the smallest k)}, \tag{30}$$

   where mixture ratio $\alpha (0 \leq \alpha \leq 1)$ for skew divergence s_α is 0.99 for example.

4. Repeat 2 and 3 until change ratio of objective function JS_W is less than small value (e.g., 10^{-8}).

The algorithm itself works well to obtain valuable clustering results. Further, if α is close to 1, skew divergence s_α becomes a good approximation of KL divergence. Therefore, restart of learning after termination with α closer to 1, such as $0.999, 0.9999, \ldots$, may lead to better clustering result.

3.2.3. Other algorithms

Slonim and Tishby [23] proposed an agglomerative hierarchical clustering algorithm, which is a hard clustering version of *Information Bottleneck algorithm* of Tishby et al. [24]. It is similar to the algorithm of Baker and McCallum [18] and merges just two clusters at every step based on the JS divergence of their distributions. A merit of the agglomerative algorithms is not affected by the difficulty of calculating KL divergence, because it just uses JS divergence. However, a merge of clusters at each step optimizes a local criterion but not a global criterion, as Dhillon et al. [8] pointed out. Therefore, clustering results may not be as good as results obtained by nonhierarchical algorithms (e.g., k-means and competitive learning) in the sense of optimizing the objective function of ITC. Additionally, hierarchical algorithms are computationally expensive, when the number of inputs is large.

Note that a lot of studies [8, 18, 23] aimed at improving accuracy of text classification using feature/word clustering based on ITC or distributional clustering. If a clustering is just a step to final goal, feature clustering is meaningful. However, features which characterize clusters should not be merged, when we aim to find clusters (topics) from a set of documents. Actually, finding topics using clustering is the aim of this chapter.

4. Evaluation of clustering

Since clustering results depend on methods (criteria and algorithms), appropriate selection of them is important. So far, we introduced two criteria for clustering. These are called **internal**

criteria that depend on their own models and not enough for evaluation. If criterion for clustering is common, we can compare clustering results by objective function of the criterion. Under a certain model that is an assumption in other word, a more probable result can be regarded as a better result. However, it is not guaranteed that the model or the assumption is reasonable at all times. Moreover, good clustering results under a certain criterion can be bad results under different criteria. A view from outside is required.

This section introduces **external criteria** that are **Purity**, **Rand index (RI)**, and **Normalized mutual information (NMI)** [25] to evaluate clustering quality and to find better clustering methods. These criteria compare clusters with a set of classes, which are produced on the basis of human judges. Here, each input data belong to one of class $A^j (j = 1,\ldots,J)$ and one of cluster $C^k (k = 1,\ldots,K)$. Let $T(C^k,A^j)$ be the number of data that belongs to both C^k and A^j.

Purity is measured by counting the number of input data from the most frequent class in each cluster. Purity can be computed as

$$\text{purity} = \frac{1}{N}\sum_{k=1}^{K} \max_j T(C^k,A^j), \tag{31}$$

where N is the total number of input data. Purity is close to 1, when each cluster has one dominant class.

Rand index (RI) checks all of the $N(N-1)/2$ pairs of input data and is defined by

$$\text{RI} = \frac{\text{a}+\text{b}}{\text{a}+\text{b}+\text{c}+\text{d}}, \tag{32}$$

where a, b, c, and d are the number of pairs in following conditions:

- "a," where the cluster number (suffix) is the same and the class number is the same
- "b," where the cluster numbers are different and the class numbers are different
- "c," where the cluster number is the same and the class numbers are different
- "d," where the cluster numbers are different and the class number is the same

The Rand index (RI) measures the percentage of agreements a+b in clusters and classes.

Normalized mutual information (NMI) is defined as

$$\text{NMI} = \frac{I(C;A)}{\big(H(C)+H(A)\big)/2}, \tag{33}$$

where, $I(C;A)$ is mutual information and $H()$ is entropy and

$$
\begin{aligned}
I(C;A) &= \sum_{k=1}^{K}\sum_{j=1}^{J} P(C^k,A^j)\log\frac{P(C^k,A^j)}{P(C^k)P(A^j)} \\
&= \sum_{k=1}^{K}\sum_{j=1}^{J} \frac{T(C^k,A^j)}{N}\log\frac{T(C^k,A^j)N}{T(C^k)T(A^j)},
\end{aligned} \tag{34}
$$

$$
H(C) = \sum_{k=1}^{K} -P(C^k)\log P(C^k) = \sum_{k=1}^{K} -\frac{T(C^k)}{N}\log\frac{T(C^k)}{N}, \tag{35}
$$

$$
H(A) = \sum_{j=1}^{J} -P(A^j)\log P(A^j) = \sum_{j=1}^{J} -\frac{T(A^j)}{N}\log\frac{T(A^j)}{N}, \tag{36}
$$

where $P(C^k)$, $P(A^j)$, and $P(C^k,A^j)$ are the probability of data being in cluster C^k, class A^j, and in the intersection of C^k and A^j, respectively. Mutual information $I(C;A)$ measures the mutual dependence between clusters C and classes A. It quantifies the amount of information obtained for classes through knowing about clusters. Hence, high NMI shows some kind of goodness about clustering in information theory.

5. Experiments

This section provides experimental results that show the effectiveness and usefulness of ITC and the proposed algorithm (sdCL: competitive learning using skew divergence). Experiments consist of two parts, experiment1 and experiment2.

In experiment1, we applied sdCL to the same data sets as used in the paper of Wang et al. [26] and compared performance of sdCL with other clustering algorithms evaluated in it. The algorithms that the paper [26] evaluated are as follows.

- **k-means (KM)**. The weights $\mathcal{W}$ are initialized by randomly [26].
- **Normalized cut (NC)** [27].
- **Maximum margin clustering (MMC)** [7].
- **Generalized maximum margin clustering (GMMC)** [28].
- **Iterative support vector regression (IterSVR)** [29].
- **Cutting plane maximum margin clustering (CPMMC)** [26], **CPM3C** [26].

As shown above, *maximum margin clustering* (MMC) [7] and related works are much focused. These works extend the idea of support vector machine (SVM) [30] to the unsupervised scenario. The experimental results obtained by the MMC technique are often better than conventional clustering methods. Among those, CPMMC and CPM3C (Cutting plane multiclass maximum margin clustering) [26] are known as successful methods. Experimental results will show that the proposed algorithm sdCL outperforms CPM3C in text data clustering.

In experiment2, we focus on text data clustering and compare performance of algorithms, sdCL, sdCLS (sdCL with splitting rule, see Sections 2.2.2 and 3.2.1), and spherical competitive learning (spCL). We also provide the evidence to support the effectiveness of ITC by detailed analysis of clustering results.

spCL is an algorithm for spherical clustering like the spherical k-means algorithm [5] that was proposed for clustering high-dimensional and sparse data, such as text data. The objective function to be maximized for the spherical clustering is cosine similarity between input vectors and the mean vector of a cluster to which they belong. To implement spCL, we turn input and weight vectors ($\boldsymbol{x}$, $\boldsymbol{w}$) into a unit vector and decide winner $\boldsymbol{w}^c$ by

$$c = \arg\max_k \cos\left(\boldsymbol{x},\boldsymbol{w}^k\right) \text{ (If there are several candidates, choose the smallest k)}, \tag{37}$$

and update the winner's weight $\boldsymbol{w}^c$ as

$$\boldsymbol{w}^c \leftarrow \frac{(1-\gamma)\boldsymbol{w}^c + \gamma\boldsymbol{x}}{\|(1-\gamma)\boldsymbol{w}^c + \gamma\boldsymbol{x}\|}. \tag{38}$$

For all competitive learning algorithms, the learning rate $\gamma = 0.01$, the number of maximum repetitions for updating weights $N_r = 1,000,000$ (termination condition), and the threshold of times for splitting rule $\theta = 1000$ are used. After competitive learning (sdCL, sdCLS, or spCL) is terminated, we apply k-means type algorithm to remove fluctuation as a post-processing. Specifically, sdKM (the k-means type algorithm using skew divergence shown in Section 3.2.2) with $\alpha = 0.999, 0.9999, 0.99999$ is applied consecutively after sdCL and sdCLS. In each learning procedure including post-processing, an operation is iterated 50 times with different initial random seeds for a given set of parameters.

5.1. Data sets

We mainly use the same data sets as used in the paper of Wang et al. [26]. When applying algorithms for ITC, we use probability distributions $P^i (i = 1,\ldots N)$ $(\mathcal{P})$ derived from original data.

1. **UCI data**. From the UCI repository,[2] we use ionosphere, digits, letter, and satellite under the same setting of the paper [26]. The digits data (8×8 matrix) are generated from bitmaps of handwritten digits. Pairs (3 vs. 8, 1 vs. 7, 2 vs. 7, and 8 vs. 9) are focused due to the difficulty of differentiating. For the letter and satellite data sets, their first two classes are used. Since the ionosphere data contain minus values and cannot be transformed to probability distributions, we do not apply ITC to them.

[2] http://archive.ics.uci.edu/ml/[Accessed: 2016-10-25].

2. **Text data**. Four text data sets: 20Newsgroups (http://qwone.com/~jason/20Newsgroups/ [Accessed: 2016-10-25]), WebKB,[3] Cora [31], and RCV1 (Reuters Corpus Volume 1) [32] are used. In experiment1, we follow the setting of the paper [26]. For 20Newsgroups data set, topic "rec" which contains four topics {autos, motorcycles, baseball, hockey} is used. From the four topics, two sets of two-class data sets {Text-1: autos vs motorcycles, Text-2: baseball vs hockey} are extracted. From WebKB data sets, the four Universities data set (Cornell, Texas, Washington, and Wisconsin University), which has seven classes (student, faculty, staff, department, course, project, and other), are used. Note that topic of the "other" class is ambiguous and may contain various topics (e.g., faculty), because it is a collection of pages that were not deemed the "main page" representing an instance of the other six classes, as pointed out in the web page of the data set. Cora data set (Cora research paper classification) [31] is a set of information of research papers classified into a topic hierarchy. From this data set, papers in subfield {data structure (DS), hardware and architecture (HA), machine learning (ML), operating system (OS), programming language (PL)} are used. We select papers that contain title and abstract. RCV1 data set contains more than 800 thousands documents to which topic category is assigned. The documents with the highest four topic codes (CCAT, ECAT, GCAT, and MCAT) in the topic codes hierarchy in the training set. Multi-labeled instances are removed.

   In experiment2, we use all of 20Newsgroups and RCV1 data sets. For RCV1 data set, we obtain 53 classes (categories) by mapping the data set to the second level of RCV1 topic hierarchy and remove multi-labeled instances. For WebKB data set, we remove "other" class due to ambiguity, use the other six classes, and do not use information of universities.

   For all text data, we remove *stop words* using *stop list* [32] and empty data, if they are not removed. In experiment1, we follow the setting of the paper [26], but properties of data sets are slightly different (see **Table 3**). For Cora data sets, the differences of data sizes are large. However, they must keep the same (or close at least) characteristics (e.g., distributions of words and topics), because they are extracted from the same source.

3. **Digits data**. USPS (16×16) and MNIST (28×28) are data sets of handwritten digits image.[4] For USPS data set, 1, 2, 3, and 4 digit images are used. For MNIST and digits data from UCI repository, all 45 pairs of digits 0–9 are used in two-class problems.

   The properties of those data sets are listed in **Table 3**.

5.2. Results of experiment1

The clustering results are shown in **Tables 4–7**, where values (except for sdCL) are the same in the paper of Wang et al. [26] (*accuracy* in that paper is equivalent to *purity* from its definition). In two-class problems, CPMMC outperforms other algorithms about purity and Rand Index (RI) in most cases. The proposed algorithm sdCL shows stable performances except for

[3]http://www.cs.cmu.edu/afs/cs.cmu.edu/project/theo-20/www/data/[Accessed: 2016-10-25].

[4]http://www.kernel-machines.org/data [Accessed: 2016-10-25].

Data	Size (N)	Feature (M)	Class (K)
Ionosphere	351	34	2
Letter	1555	16	2
Digits	1555	64	2
Satellite	2236	36	2
Text-1	1981	16,259	2
Text-2	1987	15,955	2
20Newsgroups-4	3967	24,506	4
20Newsgroups-20	18,772	60,698	20
Cora-DS	2397	5745	9
Cora-HA	913	3340	7
Cora-ML	3569	6809	7
Cora-OS	2084	5029	4
Cora-PL	3026	6069	9
WebKB-Cornell	835	5574	7
WebKB-Texas	808	4482	7
WebKB-Washington	1191	7779	7
WebKB-Wisconsin	1218	8270	7
WebKB6	4219	14,142	6
Reuters-RCV1-4	19,806	44,214	4
Reuters-RCV1-53	534,135	216,704	53
MNIST	70,000	784	2
USPS	3046	256	4

Table 3. Properties of data sets.

ionosphere to which sdCL cannot be applied. In multiclass problems, sdCL for text data (Cora, 20Newsgroups-4, and Reuters-RCV1-4) outperforms other algorithms. The results show that ITC and the proposed algorithm sdCL are effective for text data sets. Note that CPM3C shows the better results than sdCL for WebKB data. However, topic of the "other" class in WebKB is ambiguous (see Section5.1). The occupation ratio of them is large {0.710, 0.689, 0.777, 0.739} and almost same as the values of purity in CPM3C and sdCL. It means that these algorithms failed to find meaningful clusters in purity. Therefore, WebKB data are not appropriate to use for evaluation without removing "other" class.

5.3. Results of experiment2

In experiment2, we focus on text data clustering. **Table 8** shows that the proposed algorithms for ITC (sdCL and sdCLS) outperform spCL in purity, RI, and NMI. Considering that spCL is an algorithm for spherical clustering [5] which was proposed to analyze high-dimensional

Data	KM	NC	MMC	GMMC	IterSVR	CPMMC	sdCL
Iono	0.5428	0.7500	**0.7875**	0.7650	0.7770	0.7548	–
Letter	0.8206	0.7680	–	–	0.9280	**0.9502**	0.9267
Satellite	0.9593	0.9579	–	–	0.9682	**0.9879**	0.9506
Text-1	0.5053	0.9379	–	–	**0.9702**	0.9500	0.9306
Text-2	0.5038	0.9135	–	–	0.9399	**0.9721**	0.9591
Dig 3–8	0.9468	0.6500	0.9000	0.9440	0.9664	**0.9688**	0.9440
Dig 1–7	0.9445	0.5500	0.6875	0.9780	0.9945	**1.000**	**1.000**
Dig 2–7	0.9691	0.6600	0.9875	0.9950	**1.000**	**1.000**	0.9891
Dig 8–9	0.9068	0.5200	0.9625	0.8400	0.9633	**0.9812**	0.8910
UCI-dig	0.9638	0.9757	–	–	0.9818	**0.9940**	0.9516
MNIST	0.8921	0.8992	–	–	0.9241	**0.9621**	0.8812

Bold fonts indicate the maximum purities for a give data set.

Table 4. Purity comparisons for two-class problems.

Data	KM	NC	MMC	GMMC	IterSVR	CPMMC	sdCL
Iono	0.56	0.63	**0.67**	0.64	0.56	0.65	–
Letter	0.71	0.64	–	–	0.87	**0.92**	0.86
Satellite	0.92	0.92	–	–	0.94	**0.97**	0.91
Text-1	0.50	0.88	–	–	**0.94**	**0.94**	0.87
Text-2	0.50	0.84	–	–	0.89	**0.93**	0.92
Dig 3–8	0.90	0.55	0.82	0.90	0.94	**0.95**	0.89
Dig 1–7	0.99	0.50	0.57	0.96	0.99	**1.0**	**1.0**
Dig 2–7	0.94	0.55	0.98	0.99	**1.0**	**1.0**	0.98
Dig 8–9	0.84	0.50	0.93	0.73	0.93	**0.97**	0.81
UCI-dig	0.93	0.96	–	–	0.97	**0.99**	0.93
MNIST	0.81	0.82	–	–	0.86	**0.94**	0.83

Bold fonts indicate the maximum rand indices for a give data set.

Table 5. Rand Index (RI) comparisons for two-class problems.

data such as text documents, the criterion of information-theoretic clustering is worth to use for this purpose.

Table 8 also shows that sdCLS (sdCL with splitting rule, see Sections 2.2.2 and 3.2.1) is slightly better than sdCL in some cases. As far as **Figure 9** (left, right) shows, values of JS divergence for sdCLS are smaller ("better" in ITC) than sdCL, and sdCLS outperforms sdCL in purity on average. Nevertheless, an advantage of sdCLS against sdCL is not so obvious in this experiment.

Data	KM	NC	MMC	CPM3C	sdCL
UCI-digits 0689	0.4223	0.9313	0.9483	**0.9674**	0.9394
UCI-digits 1279	0.4042	0.9011	0.9191	**0.9452**	0.8300
USPS	0.9215	0.9011	0.9191	0.9452	**0.9515**
Cora-DS	0.2824	0.3688	–	0.4415	**0.5057**
Cora-HA	0.3402	0.4200	–	0.5980	**0.6145**
Cora-ML	0.2708	0.3103	–	0.4549	**0.5974**
Cora-OS	0.2387	0.2303	–	0.5916	**0.6686**
Cora-PL	0.3380	0.3397	–	0.4721	**0.4729**
WebKB-Cornell	0.5571	0.6143	–	**0.7205**	0.7192
WebKB-Texas	0.4505	0.3538	–	**0.6910**	0.6895
WebKB-Washington	0.5352	0.3285	–	**0.7817**	0.7767
WebKB-Wisconsin	0.4953	0.3331	–	**0.7425**	0.7397
20Newsgroups-4	0.3527	0.4189	–	0.7134	**0.9360**
Reuters-RCV1-4	0.2705	–	–	0.6235	**0.8064**

Bold fonts indicate the maximum purities for a give data set.

Table 6. Purity comparisons for multiclass problems.

Data	KM	NC	MMC	CPM3C	sdCL
UCI-digits 0689	0.696	0.939	0.941	**0.974**	0.945
UCI-digits 1279	0.4042	0.9011	0.9191	**0.945**	0.868
USPS	0.932	0.938	–	0.950	**0.958**
Cora-DS	0.589	0.744	–	0.746	**0.823**
Cora-HA	0.385	0.659	–	0.695	**0.767**
Cora-ML	0.514	0.720	–	0.761	**0.802**
Cora-OS	0.518	0.522	–	0.730	**0.735**
Cora-PL	0.643	0.675	–	0.712	**0.819**
WebKB-Cornell	0.603	0.602	–	**0.724**	0.483
WebKB-Texas	0.604	0.602	–	**0.712**	0.495
WebKB-Washington	0.616	0.581	–	**0.752**	0.426
WebKB-Wisconsin	0.581	0.509	–	**0.761**	0.464
20Newsgroups-4	0.581	0.496	–	0.780	**0.940**
Reuters-RCV1-4	0.471	–	–	0.703	**0.800**

Bold fonts indicate the maximum rand indices for a give data set.

Table 7. Rand Index (RI) comparisons for multiclass problems.

	sdCL			sdCLS			spCL		
Data	**Purity**	**RI**	**NMI**	**Purity**	**RI**	**NMI**	**Purity**	**RI**	**NMI**
Cora-DS	0.506	0.823	0.322	0.514	0.826	0.327	0.361	0.796	0.162
Cora-HA	0.614	0.767	0.363	0.618	0.767	0.360	0.518	0.728	0.251
Cora-ML	0.597	0.802	0.384	0.602	0.801	0.385	0.431	0.757	0.181
Cora-OS	0.669	0.735	0.313	0.685	0.741	0.325	0.582	0.673	0.187
Cora-PL	0.473	0.819	0.275	0.484	0.820	0.274	0.414	0.802	0.175
WebKB6	0.575	0.712	0.272	0.575	0.712	0.272	0.519	0.702	0.181
20News-20	0.690	0.954	0.653	0.697	0.955	0.656	0.359	0.905	0.324
RCV1-53	0.731	0.910	0.586	0.738	0.906	0.580	0.568	0.895	0.384

Table 8. Comparison for text data sets.

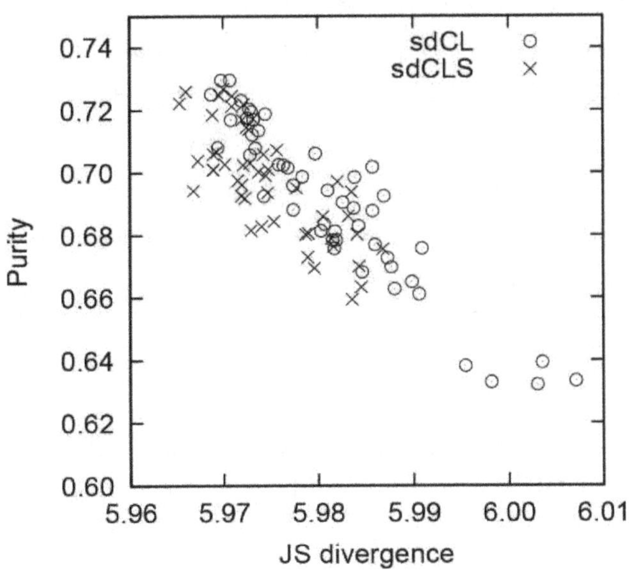

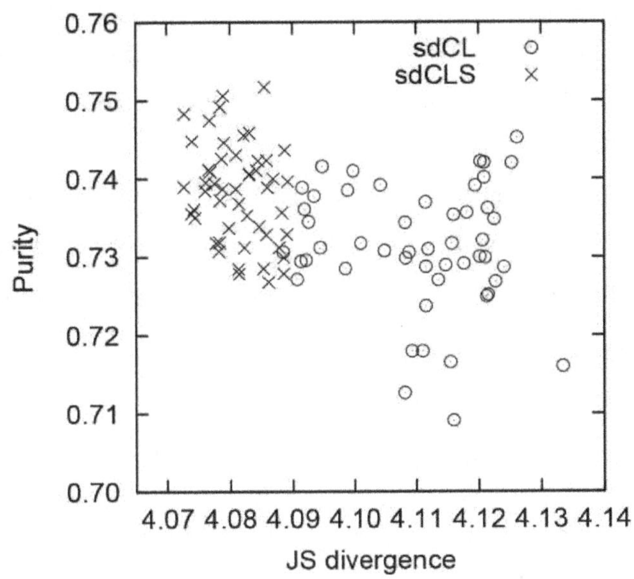

Figure 9. Purity versus JS divergence for 20Newsgroups (left) and Reuters-RCV1 (right) data sets.

Note that clustering result by dCL (competitive learning using **KL divergence**) is shown below. Since the values of NMI clearly illustrate that dCL converged to unrelated solutions to the classes, use of **skew divergence** is an effective technique to overcome this problem.

Data	Purity	RI	NMI
WebKB6	0.363	0.667	0.00629
20News-20	0.071	0.905	0.00596

alt.atheism	god writes people article atheism religion time evidence
comp.graphics	image graphics jpeg file bit images software data files ftp
comp.os.ms-windows.misc	windows file dos writes article files ms os problem win
comp.sys.ibm.pc.hardware	drive scsi card mb ide system controller bus pc writes
comp.sys.mac.hardware	mac apple writes drive system problem article mb monitor mhz
comp.windows.x	window file server windows program dos motif sun display widget
misc.forsale	sale shipping offer mail price drive condition dos st email
rec.autos	car writes article cars good engine apr ve people time
rec.motorcycles	writes bike article dod ca apr ve ride good time
rec.sport.baseball	writes year article game team baseball good games time hit
rec.sport.hockey	game team hockey writes play ca games article season year
sci.crypt	key encryption government chip writes clipper people article keys system
sci.electronics	writes article power good ve work ground time circuit ca
sci.med	writes article people medical health disease time cancer patients
sci.space	space writes nasa article earth launch orbit shuttle time system
soc.religion.christian	god people jesus church christ writes christian christians bible time
talk.politics.guns	gun people writes article guns fbi government fire time weapons
talk.politics.mideast	people israel armenian writes turkish jews article armenians israeli jewish
talk.politics.misc	people writes article president government mr stephanopoulos make time
talk.religion.misc	god writes people jesus article bible christian good christ life

Table 9. Frequent words in classes of 20Newsgroups data set.

In followings, we examine inside of clustering results obtained by ITC to make clear whether ITC helps us to find meaningful clusters and candidates of classes for classification. **Tables 9** and **10** show frequent words in classes and clusters obtained by sdCL of 20Newsgroups data set, respectively. The order of clusters is arranged so that clusters are made to correspond to classes. **Table 11** is the cross table between clusters and classes. As shown in **Table 10**, the frequent words in some clusters remind us characteristics of them to distinguish from others. For example, the words in cluster 2: "image graphics jpeg," cluster 6: "sale offer shipping," and cluster 11: "key encryption chip" remind classes (comp.graphics), (misc.forsale), and (sci.crypt), respectively. We also imagine characteristics of clusters from the words in 7, 8, 9, 10, 13, 14, and 16th clusters. These clusters have documents of one dominant class and can be regarded as candidates of classes. However, there are some exceptions. The 1st and 15th clusters have the same word "god," while classes of (alt.atheism), (soc.religion.christian), and (talk.religion.misc) have also the same word "god." The cluster 1 and class (alt.atheism) have common words "religion evidence," and the cluster 1 has many documents of the dominant class (alt.atheism). The cluster 15 and the class (soc.religion.christian) have common words "jesus bible christ church," and the cluster 15 has many documents of the dominant class (soc.religion.christian). On the other hand, there is no cluster which has

1	writes god article people religion evidence science moral objective time
2	image graphics jpeg file images ftp data bit files format
3	windows dos file system writes os files article program software
4	drive scsi mb card writes system mac article bit apple
5	window file server program motif sun widget set display output
6	sale offer shipping mail price condition st good email interested
7	car writes article cars engine good ve apr time oil
8	writes article bike dod apr ca ride ve good time
9	writes year article game team baseball good games time hit
10	game team hockey writes play ca games article year season
11	key encryption government chip writes clipper people article keys system
12	db writes article power good ground ca time ve circuit
13	writes article medical people disease health cancer patients time msg
14	space writes nasa article earth launch orbit time shuttle system
15	god jesus people bible writes christ christian church christians article
16	people writes gun article government fbi fire guns koresh batf
17	israel israeli writes jews article people arab jewish arabs state
18	people armenian turkish armenians president armenia war mr turks turkey
19	writes people article government cramer apr state make health optilink
20	mail list address email send information ca internet article writes

Table 10. Frequent words in clusters of 20Newsgroups data set.

documents of (talk.religion.misc) as dominant, and the documents of (talk.religion.misc) are mostly shared by the clusters 1 and 15. Though there is a mismatch between clusters and classes, the clustering result is also acceptable, because words in the class (talk.religion.misc) are resemble those in the class (soc.religion.christian). We can also find that cluster 4 has many documents of the two classes (comp.sys.ibm.pc.hardware) and (comp.sys.mac.hardware). From **Table 9**, those classes have similar words except for "mac" and "apple." Thus, ITC missed to detect the difference of the classes, but found the cluster with common feature of them. In this sense, the clustering result is meaningful and useful. Another example is that documents in class (talk.politics.mideast) are divided into clusters 17 and 18. It means that ITC found two topics from one class and frequency words in the clusters seem to be reasonable (see 17th and 18th clusters in **Table 10**). The characteristic of cluster 20 that has words "mail list address email send" is different from all classes as well as other clusters, but the cluster 20 has some documents in all classes (see **Table 11**). This cluster may discover that all newsgroups include documents with such words. In summary, ITC helps us to find meaningful clusters, even when clusters obtained by ITC sometimes seem not to be the same as expected classes. The detailed analysis of the clustering results above could be the evidence to support the effectiveness and usefulness of ITC.

1	625	4	9	2	2	4	2	8	0	3	2	0	5	49	12	54	1	5	13	163
2	1	632	44	19	17	101	10	7	4	8	0	14	32	12	18	2	0	0	1	1
3	0	107	707	189	58	60	27	5	0	0	1	12	40	2	3	3	0	0	1	1
4	0	74	64	630	711	18	104	3	3	1	0	2	93	6	3	2	0	1	0	1
5	1	43	53	11	15	712	0	1	0	1	1	9	2	1	1	0	1	0	0	3
6	0	9	5	18	25	2	648	17	15	3	0	0	16	1	3	1	1	0	1	0
7	0	1	2	7	8	0	35	784	65	1	1	0	25	2	7	2	2	2	1	1
8	1	3	0	0	4	6	7	36	867	2	0	2	10	6	6	0	1	1	0	2
9	4	0	4	4	0	1	4	3	4	886	18	0	2	3	2	3	0	3	6	3
10	0	0	1	1	0	1	9	3	2	37	943	0	1	1	1	0	3	2	3	0
11	2	15	2	4	6	5	5	1	0	1	2	881	36	5	6	0	20	9	15	3
12	1	11	11	52	33	2	32	14	4	3	0	8	603	28	10	2	1	0	0	1
13	1	0	1	0	5	1	3	1	1	1	2	1	5	771	5	3	0	0	2	2
14	3	9	5	2	3	5	4	4	4	1	1	1	29	17	843	2	4	0	5	5
15	94	1	2	0	0	0	2	1	1	3	0	0	0	10	2	863	5	4	6	299
16	15	2	0	5	8	2	6	36	11	0	1	18	6	8	4	7	805	2	189	93
17	9	2	1	0	1	1	0	7	0	0	0	0	2	1	8	6	2	547	14	10
18	32	0	0	0	0	0	1	1	0	2	1	1	5	3	4	12	12	333	38	12
19	3	7	8	2	16	3	12	20	4	4	7	13	6	13	11	13	33	12	470	16
20	6	50	44	33	46	58	53	35	8	34	16	27	66	48	36	22	18	19	8	11

Table 11. The number of documents about each class in each cluster.

6. Conclusion

In this chapter, we introduced information-theoretic clustering (ITC) from both theoretical and experimental side. Theoretically, we have shown the criterion, generative model, and novel algorithms for ITC. Experimentally, we showed the effectiveness and usefulness of ITC for text analysis as an important example.

A Difficulty about KL divergence

Let P and Q be a distribution whose mth random variable p_m and q_m takes the mth element of a vector $\boldsymbol{p}$ and $\boldsymbol{q}$, respectively. The Kullback-Leibler (KL) divergence to Q from P is defined to be

$$D_{\mathrm{KL}}(P\|Q) = p_m \log \frac{p_m}{q_m}. \tag{39}$$

In this definition, it is assumed that the support set of P is a subset of the support set of Q (If q_m is zero, p_m must be zero). For a given cluster C^k, there is no problem to calculate JS divergence of cluster C^k by Eq. (12), because the support set of any distribution $P^i(\in C^k)$ is the subset of the mean distribution $\overline{P}^k$. However, it is not guaranteed that KL divergence from $P^i(\in C^k)$ to $Q^t(t\neq k)$ (a weight distribution of other cluster C^t) is finite. This causes a serious problem to find similar weight distribution Q for an input distribution P. For example, lack of even one word (feature) in a distribution Q is enough not to be similar. Therefore, it is difficult to use k-means type algorithm,[5] which updates weights or clusters by batch processing, in ITC.

Acknowledgements

This work was supported by JSPS KAKENHI Grant Number 26330259.

Author details

Toshio Uchiyama

Address all correspondence to: uchiyama.toshio@do-johodai.ac.jp

Hokkaido Information University, Ebetsu-shi, Hokkaido, Japan

References

[1] A.K. Jain, M.N. Murty and P.J. Flynn, "Data clustering: a review," ACM Computing Surveys (CSUR), vol. 31, no. 3, pp. 264–323, 1999.

[2] A.K. Jain, "Data clustering: 50 years beyond k-means," Pattern Recognition Letters, vol. 31, no. 8, pp. 651–666, 2010.

[3] S. Lloyd, "Least squares quantization in pcm," IEEE Transactions on Information Theory, vol. 28, no. 2, pp. 129–137, 1982.

[4] D.E. Rumelhart and D. Zipser, "Feature discovery by competitive learning," Cognitive Science, vol. 9, pp. 75–112, 1985.

[5] I.S. Dhillon and D.S. Modha, "Concept decompositions for large sparse text data using clustering," Machine Learning, vol. 42, no. 1–2, pp. 143–175, 2001.

[6] A.Y. Ng, M.I. Jordan, Y. Weiss, et al., "On spectral clustering: analysis and an algorithm," Advances in Neural Information Processing Systems, vol. 2, pp. 849–856, 2002.

[5]k-means algorithm is known as algorithm for clustering based on the sum-of-squared-error criterion (see Section 2.2.1). Therefore we added word "type" for information-theoretic clustering.

[7] L. Xu, J. Neufeld, B. Larson, and D. Schuurmans, "Maximum margin clustering," Advances in Neural Information Processing Systems, vol. 17, pp. 1537–1544, 2004.

[8] I.S. Dhillon, S. Mallela, and R. Kumar, "A divisive information theoretic feature clustering algorithm for text classification," The Journal of Machine Learning Research, vol. 3, pp. 1265–1287, 2003.

[9] R.O. Duda and P.E. Hart, Pattern classification and scene analysis. John Wiley & Sons, New York, 1973.

[10] C.M. Bishop, Pattern recognition and machine learning (Information Science and Statistics), 1st edn. 2006. corr. 2nd printing edn. Springer, New York, 2007.

[11] J. MacQueen, et al., "Some methods for classification and analysis of multivariate observations," Proceedings of the fifth Berkeley symposium on mathematical statistics and probability, vol. 1, Oakland, CA, USA., pp. 281–297, 1967.

[12] T. Uchiyama and M.A. Arbib, "Color image segmentation using competitive learning," The IEEE Transactions on Pattern Analysis and Machine Intelligence, vol. 16, no. 12, pp. 1197–1206, 1994.

[13] Y. Linde, A. Buzo, and R.M. Gray, "An algorithm for vector quantizer design," IEEE Transactions on Communications, vol. 28, no. 1, pp. 84–95, 1980.

[14] D. Arthur and S. Vassilvitskii, "k-means++: The advantages of careful seeding," In: Proceedings of the eighteenth annual ACM-SIAM symposium on Discrete algorithms Society for Industrial and Applied Mathematics, pp. 1027–1035, 2007.

[15] S. Amari, "A theory of adaptive pattern classifiers," IEEE Transactions on Electronic Computers, vol. 16, no. 3, pp. 299–307, 1967.

[16] S.-I. Amari, "Natural gradient works efficiently in learning," Neural Computation, vol. 10, no. 2, pp. 251–276, 1998.

[17] F. Pereira, N. Tishby, and L. Lee, "Distributional clustering of english words," In: Proceedings of the 31st annual meeting on Association for Computational Linguistics Association for Computational Linguistics, pp. 183–190, 1993.

[18] L.D. Baker and A.K. McCallum, "Distributional clustering of words for text classification," In: Proceedings of the 21st annual international ACM SIGIR conference on research and development in information retrieval ACM, pp. 96–103, 1998.

[19] N. Slonim and N. Tishby, "Document clustering using word clusters via the information bottleneck method," In: Proceedings of the 23rd annual international ACM SIGIR conference on research and development in information retrieval ACM, pp. 208–215, 2000.

[20] D.M. Blei, A.Y. Ng, and M.I. Jordan, "Latent dirichlet allocation," Journal of Machine Learning Research, vol. 3, pp. 993–1022, 2003.

7

Distributions and Composite Models for Size-Type Data

Yves Dominicy and Corinne Sinner

Abstract

In the first part of this chapter, we present a sample of the best known and most used classical size distributions with their main statistical properties. In the second part, we introduce the concept of composite models and based on the size distributions of the first part, we describe those which already exist in the literature. In the last part of this chapter, we apply the described statistical size distributions and some of the composite models to two real data examples and compare their goodness-of-fit.

Keywords: size distributions, composite models, lognormal, Pareto, Weibull

1. Introduction

In statistical modeling, the continuous aim is to look for the probability law, which best describes the observations arising from a given field and which should represent the underlying data-generating process. The obtained probability distributions should possess desirable properties such as the flexibility of modeling different shapes and remain of a tractable form. This research avenue was initiated in the nineteenth century by famous mathematicians as Adolphe Quetelet, Sir Francis Galton, or Vilfredo Pareto, and since then it has never ceased. Nowadays, it still remains among the highly treated topics in statistics as shown by the large quantity of scientific papers recently published on the subject (see for instance the review papers [1] and [2]). The actual appeal for this topic is easily explained by the availability of large data sets in various scientific domains, making it essential and necessary to do further research on this subject.

In this chapter, we concentrate on probability distributions that analyze size-type data. By size distributions, we mean probability laws designed to model data that only take positive values. Positive observations appear naturally in different fields: survival analysis [3, 4], environmental science [5], network traffic modeling [6], economics [7, 8], hydrology [9], and actuarial science [10]. Given the range of various domains of application, there exists a plethora of different size distributions and it is still a very active research area [11, 12].

The structure of this chapter is as follows. In Section 2, we review the most used and well-known size distributions, and state their main statistical properties. Section 3 introduces the notion of composite models and gives a small review of the composite models in the literature, based on the size distributions depicted in Section 2. In Section 4, we apply the described size distributions of Section 2 and some of the composite models of Section 3 to two real data sets, namely, an insurance data set and an Internet traffic data set. Finally, Section 5 concludes.

2. Review of size distributions

We describe here a sample of the best known and most used size distributions. We will state their probability density function (p.d.f) and their cumulative density function (c.d.f), show their moments and their quantile function, and give the estimators obtained via maximum likelihood estimation. More specifically, we take a closer look at the lognormal, Pareto, generalized Lomax, and generalized extreme value distributions.

2.1. The lognormal distribution

The English statistician Sir Francis Galton stated that in some situations it was preferable to measure the location of a distribution with the geometric mean instead of the arithmetic mean [13]. Indeed, laws of nature often behave in multiplicative ways; thus, the geometric mean becomes more appropriate as a measure of central tendency than the arithmetic mean. As a reply to Galton's request, the Scottish physician Donald McAlister established in 1879 a theory of the exponentiated (or multiplicative) normal distribution [14], this became to be known as the lognormal distribution.

Let X be a positive random variable (r.v.) such that $\log X =^{D} Y$ is normally distributed with parameters $\mu \in \mathbb{R}$ and $\sigma > 0$. The r.v. X then has a lognormal distribution, $X \sim LN(\mu, \sigma^2)$, with probability density function (p.d.f.)

$$f(x;\mu,\sigma^2) = \frac{1}{x\sqrt{2\pi}\sigma} e^{-\frac{(\log x-\mu)^2}{2\sigma^2}}, \; x > 0. \tag{1}$$

The location parameter $\mu \in \mathbb{R}$ and the scale parameter $\sigma > 0$ are characteristic for the r.v. $\log X$. However, by the exponential transformation, the geometric mean e^{μ} becomes a scale parameter, as depicted in **Figure 1**, and the multiplicative standard deviation e^{σ} appears as shape parameter impacting the skewness (see **Figure 2**).

If random variability enjoys multiplicative effects, as stated by Galton, then a lognormal distribution must be the result. This establishes the basis of the multiplicative central limit theorem, which asserts that the geometric means of nonlognormal random variables are approximated by a lognormal distribution.

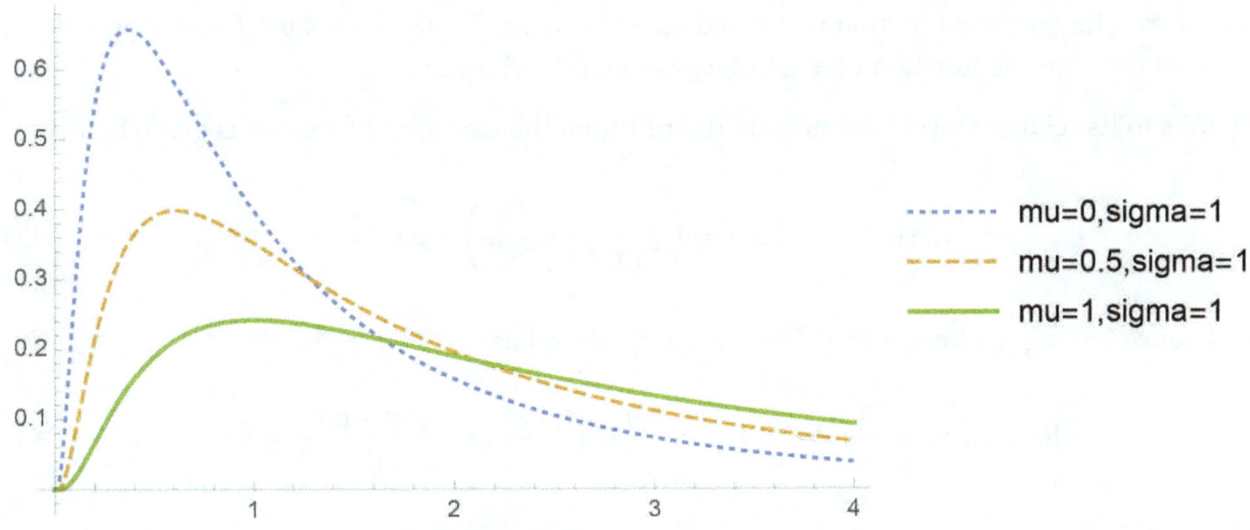

Figure 1. Density plots of the lognormal distribution with varying location parameter μ and fixed scale parameter σ.

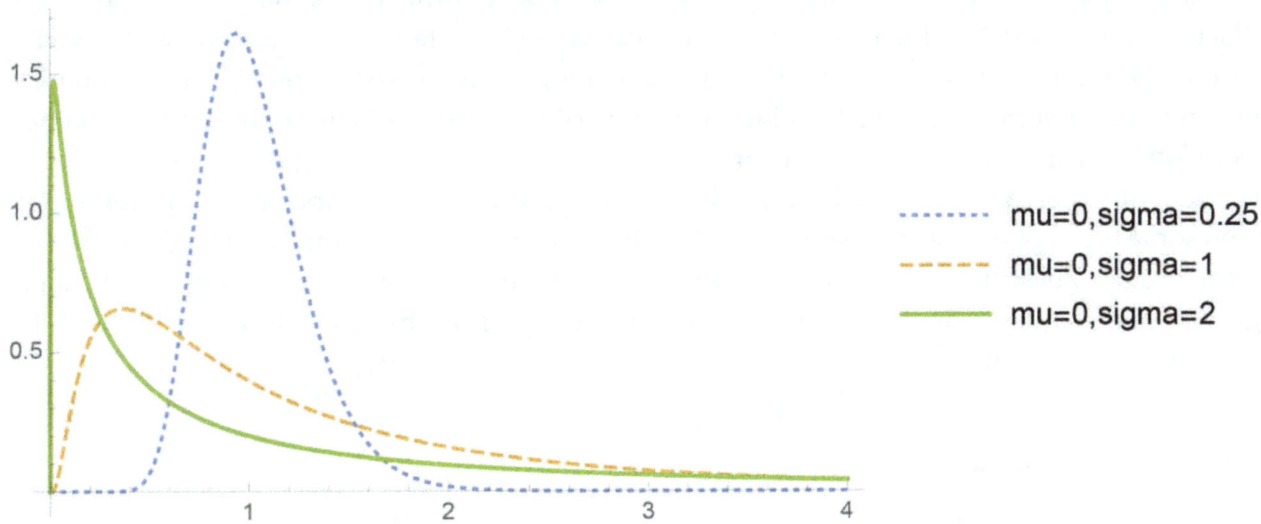

Figure 2. Density plots of the lognormal distribution with varying scale parameter σ and fixed location parameter μ.

The cumulative distribution function (c.d.f.) of the lognormal law is related to the c.d.f. of the normal distribution:

$$F(x;\mu,\sigma^2)=\Phi\left(\frac{\log x-\mu}{\sigma}\right),\ x>0, \tag{2}$$

where $\Phi(.)$ represents the c.d.f. of a standard normal distribution.

The moments of order r are conveniently expressed as $E(X^r)=e^{r\mu+\frac{r^2\sigma^2}{2}}$. Hence, the mean is given by $E(X)=e^{\mu+\frac{\sigma^2}{2}}$, and the variance by $V(X)=e^{2\mu+\sigma^2}(e^{\sigma^2}-1)$. The lognormal is a unimodal distribution and the unique mode is reached at $x_{mode}=e^{\mu-\sigma^2}$. By comparing the mean and the mode, we note that for a fixed μ, an increasing σ shifts the mode toward zero while the mean increases. The quantile function is defined as $F^{-1}(y) = e^{\mu+\sigma\Phi^{-1}(y)}$, for $0 < y < 1$ and where $\Phi^{-1}(.)$ denotes the quantile function of a standard normal distribution.

Thanks to its relationship to the normal distribution, the likelihood function is given by

$$L(x_1, \ldots, x_n|\mu, \sigma^2) = \left(\prod_{i=1}^{n} \frac{1}{x_i}\right)\left(\frac{1}{\sqrt{2\pi}\sigma}\right)^n e^{-\sum_{i=1}^{n}\frac{(\log x_i - \mu)^2}{2\sigma^2}}, \tag{3}$$

and hence the log-likelihood function can be expressed as

$$l(x_1, \ldots, x_n|\mu, \sigma^2) = -\sum_{i=1}^{n} \log x_i - \frac{n}{2}\log 2\pi - n \log \sigma - \sum_{i=1}^{n} \frac{(\log x_i - \mu)^2}{2\sigma^2}. \tag{4}$$

The maximum likelihood estimators for the mean and the scale are given by $\hat{\mu} = \frac{1}{n}\sum_{i=1}^{n}\log x_i$ and $\hat{\sigma}^2 = \frac{1}{n}\sum_{i=1}^{n}(\log x_i - \hat{\mu})^2$, respectively.

The lognormal distribution is widely used to describe natural phenomena. In finance, the Black-Scholes model, which is a mathematical model containing derivative instruments, assumes the underlying derivative price to have a lognormal distribution [15]. In economics, income data are often modeled by a lognormal distribution [16], which can be easily explained as follows: a very low percentage of earners have very low income. To gain averaged revenue is frequent, whereas an elevated income is rare. In actuarial sciences, the law is assumed to fit well some types of insurance losses [17, 18]. In 1931, the French economist and engineer Robert Pierre Louis Gibrat stated that the firm size follows a lognormal distribution as its proportional growth rate is independent of its absolute size. Other applications can be found in biology [19, 20] or in linguistics to model the number of words in a sentence [21].

2.2. The Pareto distribution

The Italian economist and engineer Vilfredo Pareto observed in 1896 that in many populations the power law $cx^{-\alpha}$, for some constant $c > 0$ and some exponent $\alpha > 0$, was an appropriate approximation of the number of individuals with income exceeding a given threshold x_0 (see for instance [22, 23]). These power laws assume that small values of x are very frequent, while large occurrences are extremely rare. Their form implies that all power laws with a particular scaling exponent are equivalent up to constant factors since each is simply a scaled version of the others. This produces the linear relationship when logarithms are taken of both $f(x)$ and x, which denotes the signature of power laws. Such distributions are known as Pareto-type distributions.

The p.d.f. of a r.v. X having a Pareto (type I) distribution with parameters $\alpha > 0$ and $x_0 > 0$ is given by

$$f(x; \alpha, x_0) = \frac{\alpha}{x_0}\left(\frac{x_0}{x}\right)^{\alpha+1}, \quad x \geq x_0. \tag{5}$$

The location parameter x_0 represents the lower bound of the data set and the shape parameter α is called the tail index or as well the Pareto index, and hence regulates the tail as can be seen in **Figure 3**. Note that a decreasing value of α implies a heavier tail.

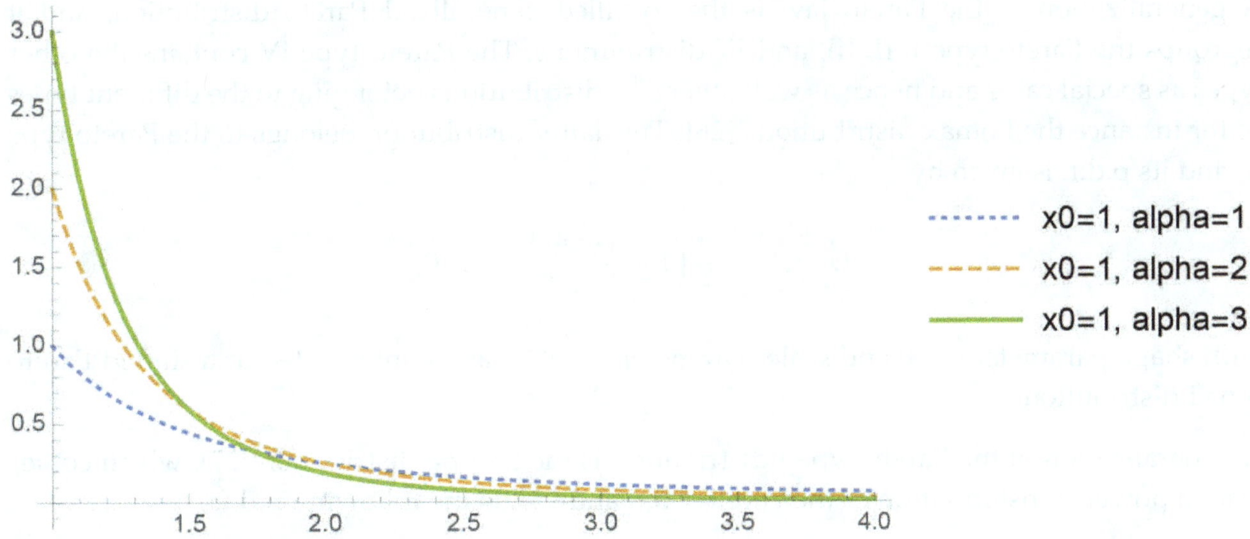

Figure 3. Density plots of the Pareto distribution with varying shape parameter α and fixed location parameter $x_0 = 1$.

The c.d.f. of the Pareto law is given by

$$F(x;\alpha,x_0) = 1-\left(\frac{x_0}{x}\right)^{\alpha},\ x \geq x_0. \tag{6}$$

For $\alpha > r$, the r-th moment of the Pareto distribution is given by $E(X^r) = \frac{\alpha x_0^r}{\alpha-r}$. The mean and the variance are then, respectively, $E(X) = \frac{\alpha x_0}{\alpha-1}$ for $\alpha > 1$ and $V(X) = \frac{\alpha x_0^2}{(\alpha-1)^2(\alpha-2)}$ for $\alpha > 2$. The quantile function is expressed as $F^{-1}(y) = \frac{x_0}{(1-y)^{\frac{1}{\alpha}}}$, for $0 < y < 1$. Being an unimodal law, the Pareto distribution reaches its peak at $x_{mode} = x_0$. As x_0 represents the minimum value of x, its estimation is straightforward: $\hat{x}_0 = \min_{i=1,\ldots,n} x_i$. The likelihood function is given by

$$L(x_1,\ldots,x_n|\alpha,x_0) = \alpha^n x_0^{n\alpha} \prod_{i=1}^{n}\left(\frac{1}{x_i}\right)^{\alpha+1}$$

and to estimate the parameter α, we maximize the following log-likelihood function

$$l(x_1,\ldots,x_n|\alpha,x_0) = n \log \alpha + n\alpha \log x_0 - (\alpha+1)\sum_{i=1}^{n} \log x_i \tag{7}$$

which yields the maximum likelihood estimator $\hat{\alpha} = \frac{n}{\sum_{i=1}^{n} \log \frac{x_i}{x_0}}$. Let us note that the maximum likelihood estimator of the tail index α corresponds to the popular Hill estimator [24], which is an estimator for the extreme value index in the extreme value theory. For a review on the Hill estimator, we refer the interested reader to reference [25]. Let us note that often the focus lies more on the power law probability distribution, which is a distribution whose density has approximately the form $L(x)x^{-\alpha}$, where $\alpha > 1$ and $L(x)$ is a slowly varying function. In many situations, it is convenient to assume a lower bound x_0 from which the law holds. Combining those two cases yields the Pareto-type distributions, or as well known in extreme value theory as distributions with regularly varying tails.

A generalization of the Pareto law is the so-called generalized Pareto distribution, and it regroups the Pareto type I, II, III, and IV distributions. The Pareto type IV contains the other types as special cases and hence as well other size distributions belonging to the different types as for instance the Lomax distributions [26]. This latter distribution belongs to the Pareto type II, and its p.d.f. is given by

$$f(x;\alpha,k) = \frac{\alpha}{k}\left(1+\frac{x}{k}\right)^{-(\alpha+1)}, \ x > 0, \tag{8}$$

with shape parameter $\alpha > 0$ and scale parameter $k > 0$. It can be interpreted as a shifted Pareto type I distribution.

A generalization of the Pareto type I distribution is the Stoppa distribution [27], which comes from a power transformation of the Pareto c.d.f. and yields the following p.d.f.

$$f(x;\alpha,\delta,x_0) = \delta\alpha x_0^{\alpha} x^{-(\alpha+1)}\left(1-\left(\frac{x}{x_0}\right)^{-\alpha}\right)^{\delta-1}, \ x > x_0, \tag{9}$$

with shape parameters $\alpha > 0$, $\delta > 0$ and location parameter $x_0 > 0$. If $\delta = 1$, we get the Pareto type I distribution. However, if the shape parameter $\delta > 1$, the Stoppa distribution presents a heavier tail than the classical Pareto law.

The Pareto distribution is often used to model fire losses in actuarial sciences [28, 29] as well as in reinsurance to approximate large losses. Originally, it was used to describe the income distribution and the allocation of wealth [22], but nowadays it is also used to model, for instance, areas burnt in forest fires or the file sizes of Internet traffic data [30]. Note that, in general, in empirical applications, the Pareto distribution does not fit for all the values but rather is used to fit their upper tail, i.e., large values. Hence, in order to fit a distribution to all the values, one often uses a composite model (see Section 3) which combines two distributions where one of both is the Pareto law.

2.3. The generalized Lomax distribution

The generalized Lomax (GL) distribution, also known as the exponentiated Lomax distribution, was introduced by Abdul-Moniem and Abdel-Hameed in 2012 [31] by powering the c.d.f. of the Lomax distribution to a positive real number.

The p.d.f. of a r.v. X following a generalized Lomax distribution with parameters $a > 0$, $b > 0$, and $k > 0$ corresponds to

$$f(x;a,b,k) = \frac{ab}{k}\left(1+\frac{x}{k}\right)^{-(a+1)}\left(1-\left(1+\frac{x}{k}\right)^{-a}\right)^{b-1}, \ x > 0. \tag{10}$$

The shape parameter a regulates the heaviness of the tail, as can be seen in **Figure 4** and the shape parameter b controls the skewness (see **Figure 5**). The parameter k is a scale parameter as depicted in **Figure 6**.

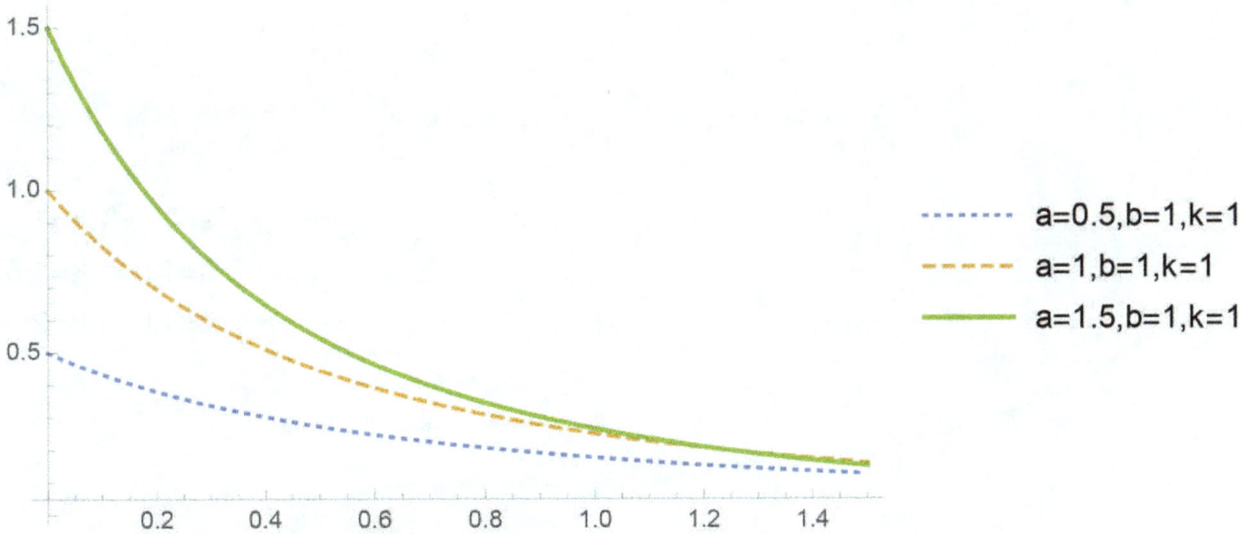

Figure 4. Density plots of the GL distribution with varying shape parameter a and fixed shape parameter b and scale parameter k.

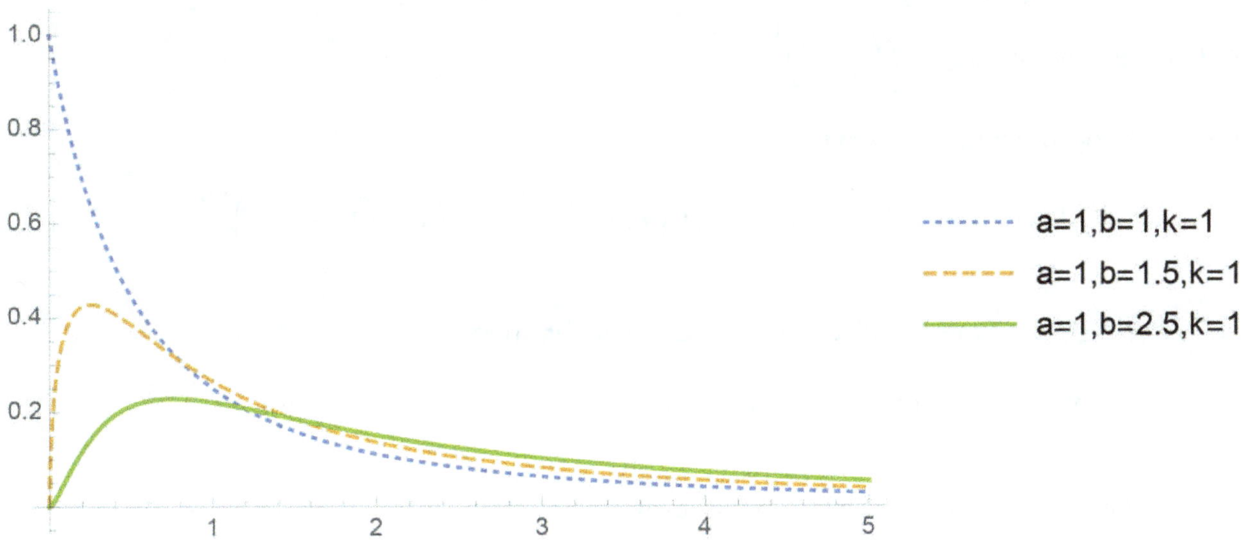

Figure 5. Density plots of the GL distribution with varying shape parameter b and fixed shape parameter a and scale parameter k.

The c.d.f. is expressed as

$$F(x;a,b,k) = \left(1-\left(1+\frac{x}{k}\right)^{-a}\right)^{b}, \; x > 0. \tag{11}$$

The moments of order r are given by $E(X^r) = bk^r\sum_{i=1}^{r}\binom{r}{i}(-1)^i B\left(1-\frac{1}{a}(r-i), b\right)$, yielding $E(X) = bkB\left(1-\frac{1}{a}, b\right)-k$ for the mean and $V(X) = bk^2\left(B\left(1-\frac{2}{a}, b\right) - b\left(B\left(1-\frac{1}{a}, b\right)\right)^2\right)$ for the

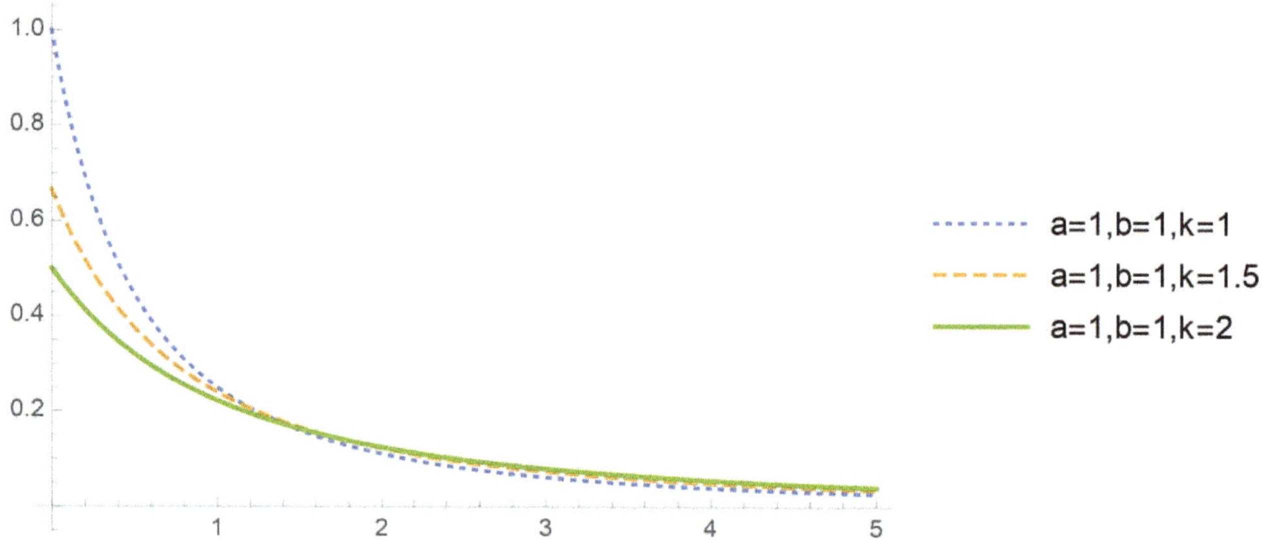

Figure 6. Density plots of the GL distribution with varying scale parameter k and fixed shape parameters a and b.

variance. The inverse c.d.f. is given by $F^{-1}(y) = k\left(\left(1 - y^{\frac{1}{b}}\right)^{\frac{-1}{a}} - 1\right)$, for $0 < y < 1$, and the unique mode is reached at $x_{mode} = k\left(\left(\frac{ab+1}{a+1}\right)^{\frac{1}{a}} - 1\right)$.

The likelihood function is given by

$$L(x_1, \ldots, x_n|a, b, k) = \left(\frac{ab}{k}\right)^n \prod_{i=1}^{n} \left(1 + \frac{x_i}{k}\right)^{-(a+1)} \prod_{i=1}^{n} \left(1 - \left(1 + \frac{x_i}{k}\right)^{-a}\right)^{b-1} \tag{12}$$

and hence the following log-likelihood function is obtained

$$l(x_1, \ldots, x_n|a, b, k) = n \log\frac{b}{k} - (a+1) \sum_{i=1}^{n} \log\left(1 + \frac{x_i}{k}\right) + (b-1) \sum_{i=1}^{n} \log\left(1 - \left(1 + \frac{x_i}{k}\right)^{-a}\right). \tag{13}$$

The calculated score functions are expressed by

$$\frac{\partial l(x_1, \ldots, x_n|a, b, k)}{\partial a} = \frac{n}{a} - \sum_{i=1}^{n} \log(1 + \frac{x_i}{k}) - a(b-1) \sum_{i=1}^{n} \frac{(1 + \frac{x_i}{k})^{-a} \log(1 + \frac{x_i}{k})}{1 - (1 + \frac{x_i}{k})^{-a}},$$

$$\frac{\partial l(x_1, \ldots, x_n|a, b, k)}{\partial b} = \frac{n}{b} + \sum_{i=1}^{n} \log\left(1 - \left(1 + \frac{x_i}{k}\right)^{-a}\right), \tag{14}$$

and

$$\frac{\partial l(x_1, \ldots, x_n|a, b, k)}{\partial k} = -\frac{n}{k} + \frac{a+1}{k^2} \sum_{i=1}^{n} \frac{x_i}{1 + \frac{x_i}{k}} - \frac{a(b-1)}{k^2} \sum_{i=1}^{n} \frac{(1 + \frac{x_i}{k})^{-(a+1)} x_i}{1 - (1 + \frac{x_i}{k})^{-a}}, \tag{15}$$

which have to be solved numerically by equating them to 0 in order to find the estimated parameters.

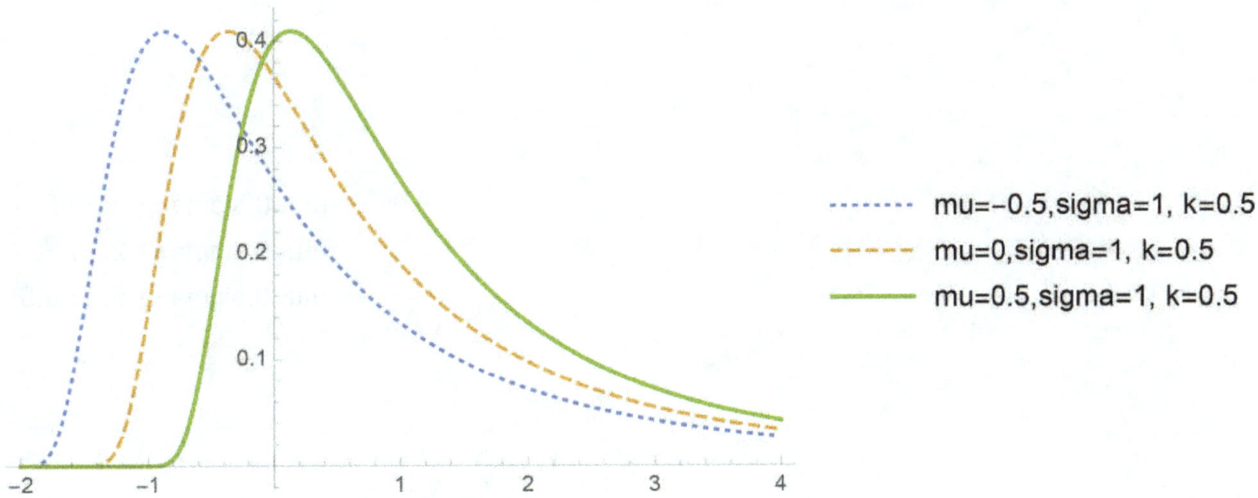

Figure 7. Density plots of the GEV distribution with varying location parameter μ.

The Lomax distribution is used to model income data, wealth allocation, and actuarial claim sizes [10]. The GL distribution is used to measure the breaking stress of carbon fibers [32], the survival times of patients getting chemotherapy treatment [33], and the number of successive failure of the air-conditioning system in airplanes [33].

2.4. The generalized extreme value distribution

The generalized extreme value (GEV) distribution is well known in extreme value theory as it combines the Gumbel, Fréchet, and Weibull distributions, which are also known as type I, II, and III extreme value distributions. The GEV distribution is the only possible limit distribution of properly normalized maxima of a sequence of independent and identically distributed random variables and this result arises from the central limit theorem of Fisher and Tippett [34]. Therefore, the GEV distribution is also known as Fisher-Tippett distribution in extreme value theory.

The GEV distribution has p.d.f.

$$f(x;\mu,\sigma,k)=\frac{1}{\sigma}\left(1+k\left(\frac{x-\mu}{\sigma}\right)\right)^{-1-\frac{1}{k}}e^{-\left(1+k\left(\frac{x-\mu}{\sigma}\right)\right)^{-\frac{1}{k}}}, \tag{16}$$

if $1+k\frac{x-\mu}{\sigma}>0$, with location parameter $\mu\in\mathbb{R}$ (see **Figure 7**), scale parameter $\sigma>0$ (see **Figure 8**), and shape parameter $k\in\mathbb{R}$, which governs the shape and the heaviness of the tail of the distribution, as can be seen in **Figure 9**.

Its c.d.f. is given by

$$F(x;\mu,\sigma,k)=e^{-\left(1+k\left(\frac{x-\mu}{\sigma}\right)\right)^{-\frac{1}{k}}}, \tag{17}$$

if $1+k\frac{x-\mu}{\sigma}>0$. The mean is defined as $E(X)=\mu-\frac{\sigma}{k}+\frac{\sigma}{k}\Gamma(1-k)$ and the variance as $V(X)=\frac{\sigma^2}{k^2}\left(\Gamma(1-2k)-\Gamma(1-k)^2\right)$, both expressed in terms of the Gamma function. The quantile

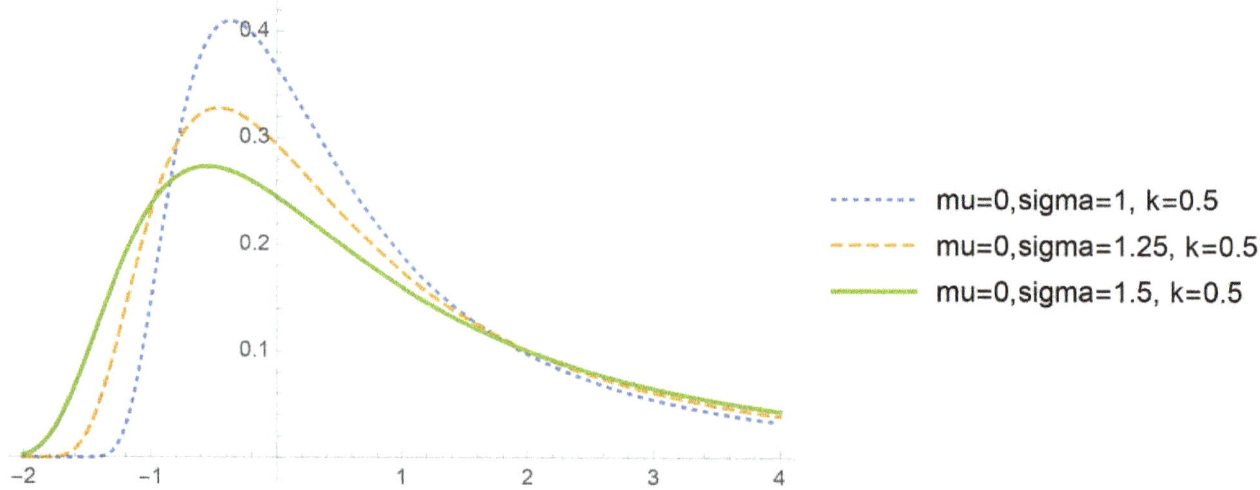

Figure 8. Density plots of the GEV distribution with varying scale parameter σ.

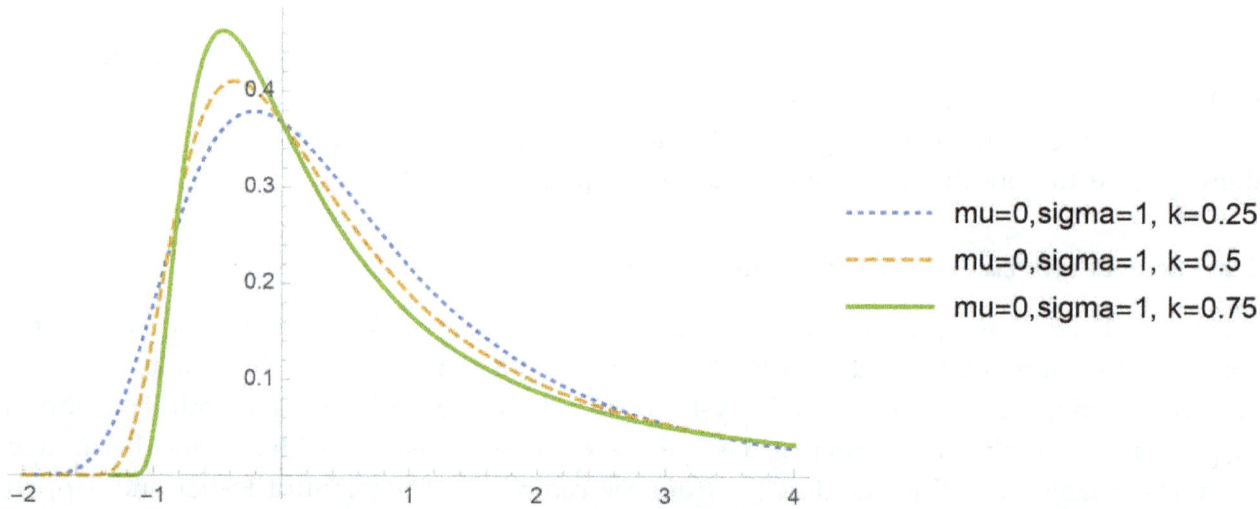

Figure 9. Density plots of the GEV distribution with varying shape parameter k.

function is given by $F^{-1}(y) = \mu + k\sigma\left((-\log y)^{-k}-1\right)$, for $0 < y < 1$ and the unique mode is reached at $x_{mode} = \mu + \frac{\sigma}{k}\left((1+k)^{-k}-1\right)$. Depending on the value of the parameter k, the GEV reduces to one of the following special cases: If $k = 0$, we obtain the Gumbel distribution, if $k > 0$, we get the Fréchet distribution, and if $k < 0$, the Weibull distribution is obtained. The parameters of the GEV distributions are estimated using the maximum likelihood approach.

We will now focus more closely on one of the three GEV distributions, namely, the Weibull distribution, which got its name from the Swedish engineer and scientist Waloddi Weibull, who analyzed it in detail in 1951. We take a look at this law as it is often used for size-type data and it is considered as an alternative to the lognormal distribution for the construction of composite models (see Section 3). The Weibull distribution belongs to power laws with an exponential cut-off; this means it is a power law multiplied by an exponential function. In these distributions, the exponential decay term overpowers the power law behavior for very large values.

The p.d.f. of the Weibull distribution is given by

$$f(x;\sigma,\tau) = \frac{\tau}{\sigma}\left(\frac{x}{\sigma}\right)^{\tau-1} e^{-\left(\frac{x}{\sigma}\right)^{\tau}},\ x \geq 0, \tag{18}$$

with shape parameter $\tau > 0$ governing the heaviness of the tail and scale parameter $\sigma > 0$.

The distribution has c.d.f.

$$F(x;\sigma,\tau) = 1 - e^{-\left(\frac{x}{\sigma}\right)^{\tau}},\ x \geq 0. \tag{19}$$

The quantile function is given by $F^{-1}(y) = \sigma(-\log(1-y))^{\frac{1}{\tau}}$, for $0 < y < 1$. The r-th moment is given by $E(X^r) = \sigma^r\Gamma(1+\frac{r}{\tau})$. Hence, the expectation and the variance are expressed as $E(X) = \sigma\Gamma(1+\frac{1}{\tau})$and $V(X) = \sigma^2\left(\Gamma(1+\frac{2}{\tau}) - \Gamma(1+\frac{1}{\tau})^2\right)$, respectively. The Weibull distribution is unimodal and for $\tau > 1$ it reaches the mode at $x_{mode} = \sigma\left(\frac{\tau-1}{\tau}\right)^{\frac{1}{\alpha}}$ and for $\tau = 1$ the mode is reached at 0.

The parameters of the Weibull distribution are estimated via the maximum likelihood method. The corresponding likelihood and log-likelihood functions are given respectively by

$$L(x_1,\ldots,x_n|\sigma,\tau) = \frac{\tau^n}{\sigma^n}\prod_{i=1}^{n}\left(\frac{x_i}{\sigma}\right)^{\tau-1} e^{-\sum\limits_{i=1}^{n}\left(\frac{x_i}{\sigma}\right)^{\tau}}, \tag{20}$$

and

$$l(x_1,\ldots,x_n|\sigma,\tau) = n\log\tau - n\,\tau\log\sigma + (\tau-1)\sum_{i=1}^{n}\log(x_i) - \sum_{i=1}^{n}\left(\frac{x_i}{\sigma}\right)^{\tau}. \tag{21}$$

The maximum likelihood estimator for the scale parameter σ, given τ, is $\hat{\sigma}^{\tau} = \frac{1}{n}\sum_{i=1}^{n}x_i^{\tau}$ and the maximum likelihood estimator for the shape parameter τ is given by an implicit function which has to be solved numerically: $\hat{\tau}^{-1} = \frac{\sum_{i=1}^{n}x_i^{\tau}\log(x_i)}{\sum_{i=1}^{n}x_i^{\tau}} - \frac{1}{n}\sum_{i=1}^{n}\log(x_i)$.

In risk management, finance, and insurance, the risk measure "Value at Risk" is assessed by considering the GEV distribution [35, 36]. They are as well used in hydrology [37, 38], telecommunications [39], and meteorology [40]. In material sciences, the Weibull is widely used thanks to its flexibility [41]. Other examples include wind speed distributions [42], forecasting technological change [43], the size of reinsurance claims [10], hydrology [9], and areas burnt in forest fires [44].

3. Composite models

Given the wealth of distinct size distributions, as can be seen from the small sample of size distributions described in the previous section, the practitioner is often confronted to the following question: Which size distribution shall he/she use in which situation? The variation in the shapes, size distributions can take, for instance between the Pareto distribution and the lognormal distribution, renders the choice very complicated in practice.

For example, insurance companies face sometimes losses, which emerge from a combination of moderate and large claims. In order to model those large losses, the Pareto distribution seems to be the size distribution favored by practitioners. However, when losses consist of smaller values with high frequencies and larger losses with low frequencies, the lognormal or the Weibull distributions are preferred [45]. Nevertheless, no classical size distribution provides an acceptable fit for both small and large losses. On one hand, the Pareto fits well the tail, but on the other hand, lognormal and Weibull distributions produce an overall good fit, but fit badly the tail.

A solution to this dilemma comes from the composite parametric models introduced in 2005 by Cooray and Ananda [46]. The idea of the composite models is to join together two weighted distributions at a given threshold value. In statistical terms, let X be a r.v. and denote by $f_1(.)$ the p.d.f. of the first distribution and by $f_2(.)$ the p.d.f. of the second distribution. Let $F_1(.)$ and $F_2(.)$ be the corresponding c.d.f., respectively. Scollnik [47] noticed that the p.d.f. of a composite model can then be expressed as

$$f(x) = \begin{cases} c f_1^*(x), & -\infty < x \leq \theta \\ (1-c) f_2^*(x), & \theta < x < \infty, \end{cases} \tag{22}$$

where c is a normalization constant in [0,1], θ represents the threshold value, $f_1^*(x) = \frac{f_1(x)}{F_1(\theta)}$ for $-\infty < x \leq \theta$, and $f_2^*(x) = \frac{f_2(x)}{1-F_2(\theta)}$ for $\theta < x < \infty$. In our setting, the considered composite models piece together two different size distributions with different shapes and tail-weights at a specific threshold. As size distributions are only for positive values, the p.d.f. of a composite model is rewritten as

$$f(x) = \begin{cases} c f_1^*(x), & 0 < x \leq \theta \\ (1-c) f_2^*(x), & \theta < x < \infty, \end{cases} \tag{23}$$

where $0 \leq c \leq 1$. The composite model can as well be interpreted as a two-component mixture model with mixing weights c and $(1 - c)$. Hence, it can be seen as a convex sum of two density functions $f(x) = c\, f_1^*(x) + (1 - c)\, f_2^*(x)$, as noted in [47].

As we have a threshold that cuts the composite model distribution into two, from a mathematical point of view, we need continuity and differentiability conditions at the threshold to yield a smooth density function. In order to make $f(x)$ continuous, the following condition $f(\theta^-) = f(\theta^+)$ is imposed and yields

$$c = \frac{f_2(\theta) F_1(\theta)}{f_2(\theta) F_1(\theta) + f_1(\theta)\left(1 - F_2(\theta)\right)}. \tag{24}$$

The differential condition at the threshold value is given by $f'(\theta^-) = f'(\theta^+)$ and yields

$$c = \frac{f_2'(\theta)F_1(\theta)}{f_2'(\theta)F_1(\theta) + f_1'(\theta)\left(1 - F_2(\theta)\right)}. \tag{25}$$

If we combine the two results for the normalization constant c, we obtain the additional restriction for θ, i.e., $\frac{f_1(\theta)}{f_2(\theta)} = \frac{f_1'(\theta)}{f_2'(\theta)}$. Let us remark that in reference [48] they use a mode-matching procedure instead and state that it gives much simpler derivation of the model and allows for an easier implementation with any distribution which has a mode that has a closed form expression. Instead of having as threshold value θ, they use the modal value x_m. Denote by x_{m1} and x_{m2} the modes of the distributions used by the first and second components of the composite model, then the mode-matching conditions are $x_{m1} = x_{m2}$ and $f^*(x_{m1}) = f^*(x_{m2})$. The latter implies the continuity condition, and the former equality allows dropping the labels 1 and 2, which yields the following condition

$$c = \frac{f_2(x_m)F_1(x_m)}{f_2(x_m)F_1(x_m) + f_1(x_m)\left(1 - F_2(x_m)\right)}. \tag{26}$$

Remark that the derivative at the mode is 0, hence the differentiability condition is satisfied.

The c.d.f. of a composite model of size distributions is given by

$$F(x) = \begin{cases} c\,\dfrac{F_1(x)}{F_1(\theta)}, & 0 < x \leq \theta \\ c + (1 - c)\,\dfrac{F_2(x) - F_2(\theta)}{1 - F_2(\theta)}, & \theta < x < \infty. \end{cases} \tag{27}$$

The moments of the r-th order can be expressed using this formula $E_r(f) = cE_r(f_1^*) + (1 - c)E_r(f_2^*)$.

Statistical inference for composite models is done using the classical maximum likelihood (ML) estimation approach. The ML estimation for composite models was first presented in [46] and as well in [49]. In order to apply the ML approach, we have to know the integer value m such that the unknown threshold parameter θ is in between the m-th and $m + 1$-th observation. If we assume somehow that we know the value of the integer m, we would be able to write out explicitly the likelihood function. However, unfortunately, we do not know the exact value of m and as m changes, the ML estimation changes. Therefore, the following ML estimation algorithm was proposed where we have s parameters ρ_i for $i = 1, \ldots, s$. In a first step, for each integer $m = 1, \ldots, n - 1$ we estimate the parameters as solution of the following ML system

$$\begin{cases} \dfrac{\partial \log L}{\partial \rho_i} = 0, \; i = 1, \ldots, s, \\ \dfrac{\partial \log L}{\partial \theta} = 0. \end{cases} \tag{28}$$

If the inequality $x_m \leq \hat{\theta} \leq x_{m+1}$ holds, then the ML estimators can be denoted as $\hat{\theta}$ and $\hat{\rho}_i$ for $i = 1, \ldots, s$. However, a second step is needed in case the first step does not provide any satisfying result meaning that we are either in one of the following two settings $m = n$ or $m = 0$. This implies that the use of f_1 and f_2 are recommended for the likelihood function, respectively. For the ML procedure, one needs to check $n - 1$ intervals. Thus, the computing time strongly depends on the magnitude of n. For large n this leads to a complex system of equations that must be solved numerically.

In reference [50], the authors propose an alternative algorithm based on quantiles and a moment matching approach. In a first step, let us denote by q_1 and q_3 the first and third empirical quartiles of the data sample. We assume that $q_1 \leq \theta \leq q_3$. Then we use the method of moments to match the first $s - 1$ empirical moments with their theoretical counterparts, and we add two more equations from matching two quartiles

$$\begin{cases} c\dfrac{F_1(q_1)}{F_1(\theta)} = 0.25, \\ c + (1-c)\dfrac{F_2(q_3) - F_2(\theta)}{1 - F_2(\theta)} = 0.75. \end{cases} \tag{29}$$

If no result is obtained, we move to a second step where we assume that the first and third quartiles are smaller than the threshold θ, and proceed like in the first step except using now the following two quartiles' equations

$$\begin{cases} c\dfrac{F_1(q_1)}{F_1(\theta)} = 0.25, \\ c\dfrac{F_1(q_3)}{F_1(\theta)} = 0.75. \end{cases} \tag{30}$$

If we still have no solution, we finally assume that the first and third quartiles are greater than θ and proceed again in a similar fashion as in the first step with the two equations

$$\begin{cases} c + (1-c)\dfrac{F_2(q_1) - F_2(\theta)}{1 - F_2(\theta)} = 0.25, \\ c + (1-c)\dfrac{F_2(q_3) - F_2(\theta)}{1 - F_2(\theta)} = 0.75. \end{cases} \tag{31}$$

Let us remark that those equations have to be solved numerically. Note that once we have a solution from this quantile and moment matching procedure, we can use the ML approach explained above to improve the result as now we have some *a priori* information on the parameter θ and hence on the integer m.

In general, in the area of size distributions, composite models comprise a lognormal or Weibull distribution up to a given threshold value and some form of the Pareto distribution thereafter. The obtained models are close in shape to the lognormal or Weibull law but with a thicker tail due to the Pareto distribution, see **Figure 10** and **Figure 11**.

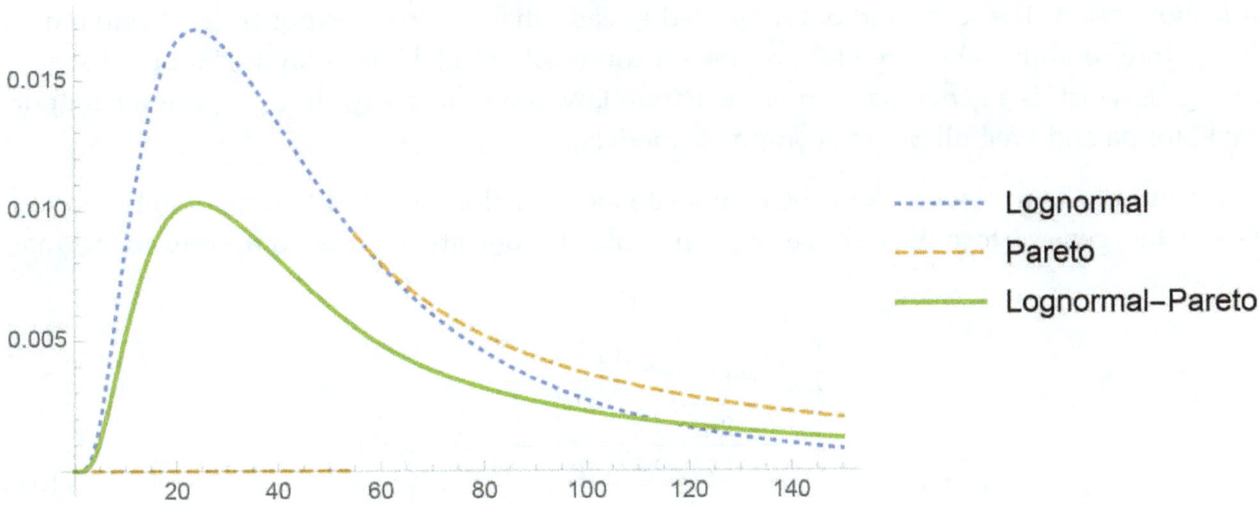

Figure 10. Density plot of the composite lognormal-Pareto model with $\theta = 0.55$ and $\alpha = 0.5$.

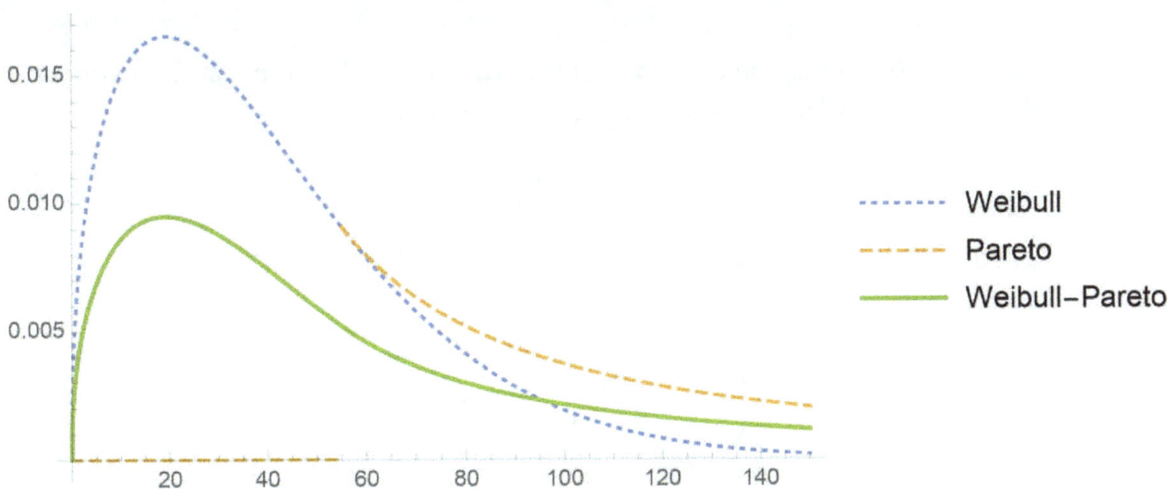

Figure 11. Density plot of the composite Weibull-Pareto model with $\theta = 0.55$ and $\tau = 1.42867$.

This research area for size distributions was initiated by Cooray and Ananda in 2005 [46], who proposed the composite lognormal-Pareto model. They suggested that this composite model may be better suited for insurers when confronted to smaller losses with high frequencies as well as for larger values with lower frequencies. The lognormal-Pareto composite model introduced in reference [46] has been further enhanced by Scollnik [47]. In that paper, the author noticed that the two-component composite model is very restrictive since it has fixed and *a priori* known mixing weights. Hence, he improved the model by using unrestricted mixing weights as coefficients in each component. In a similar way, the article [51] improves the composite Weibull-Pareto model proposed by reference [52]. Those are the composite models that will be described in more detail in the sequel. The papers [47] and [51] consider beside the classical Pareto distribution as well the Pareto type II distribution, known also as the Lomax distribution, as an alternative above the threshold value. In 2013, Teodorescu and Vernic [50] replace the lognormal distribution by any arbitrary continuous distribution, and they analyze in detail the composite Weibull-Pareto and the composite Gamma-Pareto models, and use as well the Lomax distribution as an alternative to the Pareto distribution above the threshold point. The same authors suggested already the composite exponential-Pareto model [50]. More recently, reference [48] proposes a composite model based on the Stoppa distribution [27], which is a generalization of the Pareto law. More precisely, they propose the lognormal-Stoppa and Weibull-Stoppa composite models.

Let us now take a closer look at the composite lognormal-Pareto and Weibull-Pareto models. Given the general formulas above we can write the density for the composite lognormal-Pareto as

$$f(x) = \begin{cases} c\,\dfrac{\dfrac{1}{x\sqrt{2\pi}\sigma}e^{-\dfrac{(\log x-\mu)^2}{2\sigma^2}}}{\Phi\left(\dfrac{\log\theta-\mu}{\sigma}\right)}, & 0 < x \le \theta \\[2ex] (1-c)\dfrac{\alpha}{\theta}\left(\dfrac{\theta}{x}\right)^{\alpha+1}, & \theta < x < \infty, \end{cases} \tag{32}$$

with $0 \le c \le 1$ and $\Phi(.)$ denoting the c.d.f. of a standard normal distribution. In a similar way, the p.d.f. for the composite Weibull-Pareto can be written as

$$f(x) = \begin{cases} c\,\dfrac{\dfrac{\tau}{\sigma}\left(\dfrac{x}{\sigma}\right)^{\tau-1}e^{-\left(\dfrac{x}{\sigma}\right)^{\tau}}}{1-e^{-\left(\dfrac{\theta}{\sigma}\right)^{\alpha}}}, & 0 < x \le \theta \\[2ex] (1-c)\dfrac{\alpha}{\theta}\left(\dfrac{\theta}{x}\right)^{\alpha+1}, & \theta < x < \infty. \end{cases} \tag{33}$$

with $0 \le c \le 1$.

By verifying the continuity and differentiability conditions at the threshold point θ, we obtain for the composite lognormal-Pareto model:

$$c = \frac{\frac{\alpha}{\theta}\left(\frac{\theta}{x}\right)^{\alpha+1}\Phi\left(\frac{\log\theta-\mu}{\sigma}\right)}{\frac{\alpha}{\theta}\left(\frac{\theta}{x}\right)^{\alpha+1}\Phi\left(\frac{\log\theta-\mu}{\sigma}\right)+\frac{1}{\theta\sqrt{2\pi}\sigma}e^{-\frac{(\log\theta-\mu)^2}{2\sigma^2}}} \tag{34}$$

and

$$\alpha\sigma = \frac{\log\theta-\mu}{\sigma}. \tag{35}$$

These conditions guarantee that the p.d.f. of the composite lognormal-Pareto is continuous and smooth at the threshold value θ. The continuity and differentiability conditions at θ, for the composite Weibull-Pareto, yield:

$$c = \frac{\frac{\alpha}{\theta}\left(\frac{\theta}{x}\right)^{\alpha+1}\left(1 - e^{-\left(\frac{\theta}{\sigma}\right)^{\tau}}\right)}{\frac{\alpha}{\theta}\left(\frac{\theta}{x}\right)^{\alpha+1}\left(1 - e^{-\left(\frac{\theta}{\sigma}\right)^{\tau}}\right) + \frac{\tau}{\sigma}\left(\frac{\theta}{\sigma}\right)^{\tau-1} e^{-\left(\frac{\theta}{\sigma}\right)^{\tau}}} \tag{36}$$

and

$$\left(\frac{\theta}{\sigma}\right)^{\tau} = \frac{\alpha}{\tau} + 1. \tag{37}$$

These conditions guarantee the continuity and smoothness of the p.d.f. of the composite Weibull-Pareto at the threshold point θ.

The c.d.f. of the composite lognormal-Pareto and Weibull-Pareto is given, respectively, by

$$F(x) = \begin{cases} c\dfrac{\Phi\left(\dfrac{\log x - \mu}{\sigma}\right)}{\Phi\left(\dfrac{\log\theta - \mu}{\sigma}\right)}, & 0 < x \leq \theta \\ c + (1-c)\left(1 - \left(\frac{\theta}{x}\right)^{\alpha}\right), & \theta < x < \infty. \end{cases} \tag{38}$$

and

$$F(x) = \begin{cases} c\dfrac{1 - e^{-\left(\frac{x}{\sigma}\right)^{\tau}}}{1 - e^{-\left(\frac{\theta}{\sigma}\right)^{\tau}}}, & 0 < x \leq \theta \\ c + (1-c)\left(1 - \left(\frac{\theta}{x}\right)^{\alpha}\right), & \theta < x < \infty. \end{cases} \tag{39}$$

Finally, the moments of order r of the composite lognormal-Pareto and Weibull-Pareto are given by

$$E(X^r) = c\frac{e^{r\mu + \frac{r^2\sigma^2}{2}}}{\Phi\left(\frac{\log\theta - \mu}{\sigma}\right)} + (1-c)\frac{\alpha\theta^r}{\alpha - r} \tag{40}$$

and

$$E(X^r) = c\frac{\sigma^r \Gamma\left(\frac{r}{\tau} + 1; \left(\frac{\theta}{\sigma}\right)^{\tau}\right)}{1 - e^{-\left(\frac{\theta}{\sigma}\right)^{\tau}}} + (1-c)\frac{\alpha\theta^r}{\alpha - r} \tag{41}$$

for $\tau > r$, respectively.

To estimate the composite lognormal-Pareto and the composite Weibull-Pareto models, the algorithms described above are used.

4. Applications

In this section, we focus on two applications to real data sets, one from actuarial sciences, dealing with fire losses and one on Internet traffic data. We will analyze these two data sets with the size distributions seen in Section 2 and the two composite models, namely, the lognormal-Pareto and the Weibull-Pareto, seen in Section 3. In order to compare the distributions, we used the following three criteria:

1. The maximum log-likelihood (MLL) value: the larger the value, the better the fit of the distribution to the data set.
2. The Akaike information criterion (AIC):

$$\text{AIC} = 2p–2MLL,$$

where p represents the number of parameters to estimate. This criterion represents a measure of the relative quality of a distribution given a set of laws. The distribution with the lowest AIC value is preferred.

3. The Bayesian information criterion (BIC):

$$\text{BIC} = p\log n–2MLL,$$

where n represents the length of the data set and p the number of parameters to estimate. This criterion is used to choose a distribution among a finite set of laws. The distribution with the lowest BIC is preferred.

The AIC and BIC give a trade-off between a reward for a good goodness-of-fit performance and a penalty for an increasing number of parameters to estimate. The BIC tends to favor more parsimonious models than does the AIC.

We carried out the calculations with *Wolfram Mathematica 10*. To calculate the MLL, AIC, and BIC values for the size distributions of Section 2, we used the function *NMaximize* with numerical maximization algorithm *Random Search method* enhanced with the option *Interior-Point*. For the composite lognormal-Pareto and the composite Weibull-Pareto models, we used the estimation algorithms described in Section 3.

4.1. Danish fire losses

In this example, we analyze a classical insurance data set. This is the set of Danish data on 2492 fire insurance losses in Danish Krone (DK) from the years 1980 to 1990 inclusive. The data set can be found in the "SMPracticals" add-on package for R, available from the CRAN website cran.r-project.org.

The comparison of the considered distributions using the three criteria explained above is presented in **Table 1**. The estimated values for the fitted distributions are given in **Table 2**.

Distribution	p	MLL	AIC	BIC
Lognormal	2	-4433.89	8871.78	8883.42
Pareto	2	–5675.09	11354.20	11365.80
GL	3	–3967.18	7940.36	7957.82
GEV	3	–3955.43	7916.86	7934.32
Lognormal-Pareto	2	–3877.84	7759.68	7771.32
Weibull-Pareto	2	–3959.78	7923.56	7935.20

Table 1. MLL, AIC, and BIC values for the Danish fire data set.

Distribution			
Lognormal	$\widehat{\mu} = 0.671854$	$\hat{\sigma} = 0.732317$	
Pareto	$\hat{\alpha} = 0.545817$	$\hat{x}_0 = 0.313404$	
GL	$\hat{\alpha} = 2.01251$	$\hat{b} = 435198$	$\hat{k} = 0.00227572$
GEV	$\widehat{\mu} = 1.42575$	$\hat{\sigma} = 0.712043$	$\hat{k} = 0.545094$
Lognormal-Pareto	$\hat{\alpha} = 1.43633$	$\hat{\theta} = 1.38513$	
Weibull-Pareto	$\hat{\tau} = 4.43613$	$\hat{\theta} = 1.46597$	

Table 2. Estimated values for the fitted distributions for the Danish fire data set.

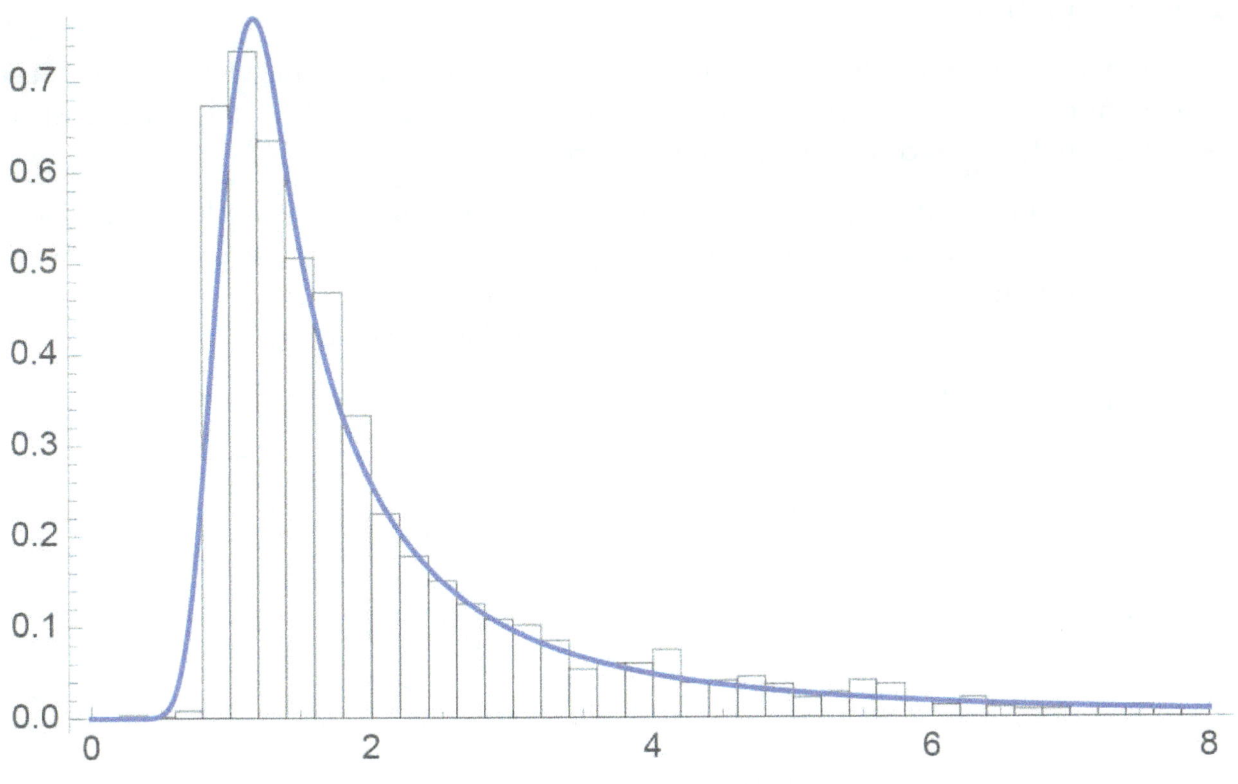

Figure 12. Histogram of the Danish fire data with the fitted density of the composite lognormal-Pareto model.

Distribution	p	MLL	AIC	BIC
Lognormal	2	–39582.2	79168.4	79180.5
Pareto	2	–43031.7	86067.4	86079.5
GL	3	–39581.7	79169.4	79187.6
GEV	3	–39608.4	79222.8	79241.0
Lognormal-Pareto	2	–40098.4	80200.8	80212.9
Weibull-Pareto	2	–42823.9	85651.8	85663.9

Table 3. MLL, AIC, and BIC values for the Internet traffic data.

With a MLL value of -3877.84 and only two parameters, yielding the values AIC = 7759.68 and BIC = 7771.32, the lognormal-Pareto model provides a better fit than the other models for the given data set. A visual conclusion of the fit can be seen in **Figure 12**.

As the data present a humped shape behavior for the lower values and tail behavior for the upper values; this example justifies the use and the necessity of the composite lognormal-Pareto model.

This data set has also been analyzed in reference [46] where the composite lognormal-Pareto model was introduced and reference [51] applied as well the Weibull-Pareto model to this data set. The results we obtain above coincide with their results.

4.2. Internet traffic data

In the second empirical illustration, we analyze Internet traffic data, which have already been analyzed from a Bayesian point of view in references [53] and [54]. This data set consists of 3143 transferred bytes/second within consecutive seconds.

Based on the MLL, the AIC, and BIC values represented in **Table 3**, we conclude that among the considered laws, the lognormal distribution performs the best fit, closely followed by the GL and the GEV distributions. The two considered composite models do not provide good fits for this example. The estimated values for the fitted densities are given in **Table 4**.

Figure 13 provides a visual proof of the goodness-of-fit of the lognormal distribution.

Distribution			
Lognormal	$\widehat{\mu} = 11.6518$	$\hat{\sigma} = 0.62067$	
Pareto	$\hat{\alpha} = 0.353628$	$\hat{x}_0 = 6795$	
GL	$\hat{\alpha} = 13.6735$	$\hat{b} = 4.08831$	$\hat{k} = 808429$
GEV	$\widehat{\mu} = 94465$	$\hat{\sigma} = 54467.5$	$\hat{k} = 0.204602$
Lognormal-Pareto	$\hat{\alpha} = 1.1.05077$	$\hat{\theta} = 85064.5$	
Weibull-Pareto	$\hat{\tau} = 1.12043$	$\hat{\theta} = 79366.3$	

Table 4. Estimated values for the fitted distributions for the Internet traffic data.

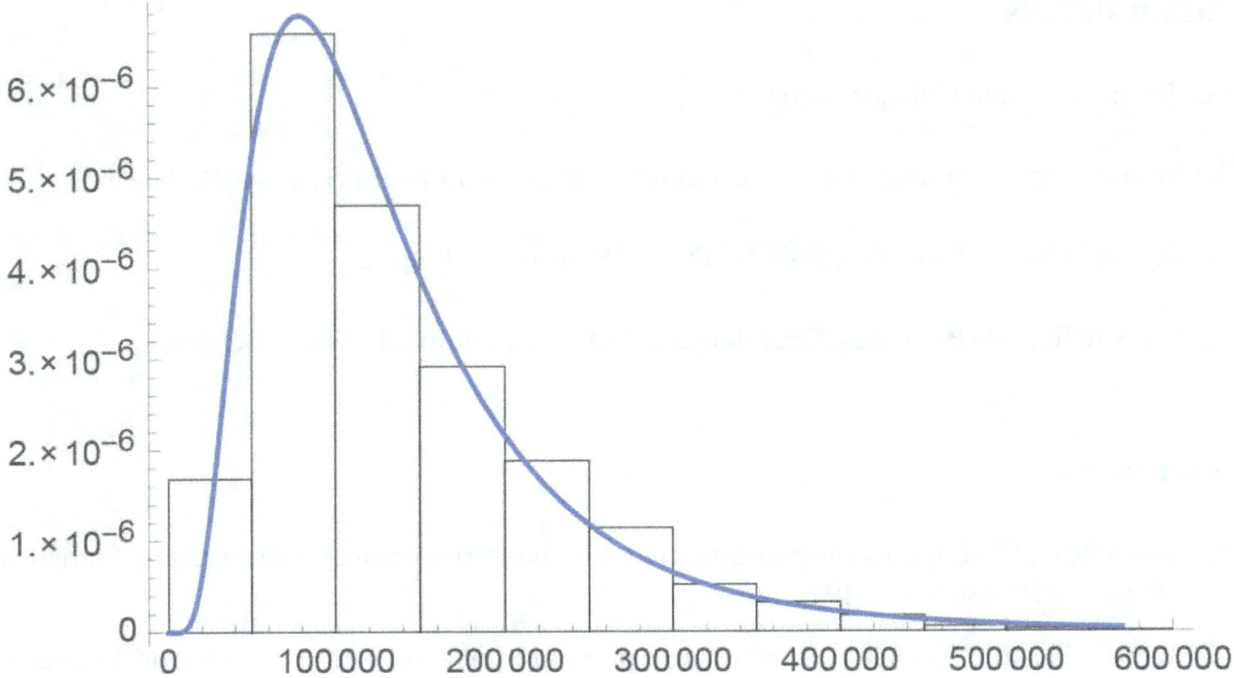

Figure 13. Histogram of the Internet traffic data with the fitted lognormal density.

5. Conclusion

To sum up, we review in this chapter the notion of size distributions by presenting the best known and most used ones. We further describe the general concept of composite models based on size distributions and present in more details the composite lognormal-Pareto and the composite Weibull-Pareto models. Besides providing their main statistical properties, we illustrate the size distributions and composite models by applying them to two real application examples to emphasize their use in practice. We compare the goodness-of-fit of the considered distributions using as criteria the MLL, AIC, and BIC. For the first data set dealing with fire losses we find that the composite lognormal-Pareto model performs the best, hinting at the usefulness of composite models in this research area. However, for the second data set on Internet traffic, the simple lognormal distribution outperforms the other size distributions and composite models. This shows how delicate the choice is for a practitioner when confronted with the question which distribution or model he/she should use on a given data set. The composite models are already quite flexible, but given the different shapes a data set can take, there is a quest for even more flexible distributions. In the literature, some families of distributions are proposed, which contain many of the classical size distributions and hence can model very diverse behaviors. The most popular one is the generalized beta distribution presented in reference [55], and very recently reference [56] introduces a new flexible distribution called the interpolating family of size distributions. Those distributions are quite flexible as they enable to model very distinct shapes and probably constitute the future avenue of research in the area of size distributions.

Author details

Yves Dominicy[1]* and Corinne Sinner[2]*

*Address all correspondence to: yves.dominicy@ulb.ac.be and corinne.sinner@ulb.ac.be

1 Université libre de Bruxelles, SBSEM, ECARES, Brussels, Belgium

2 Université libre de Bruxelles, Département de Mathématique, Brussels, Belgium

References

[1] Jones M.C. On families of distributions with shape parameters. International Statistical Review. 2015;**83**(2):175– 192.

[2] Ley C. Flexible modelling in statistics: past, present and future. Journal de la Société Française de Statistique. 2015;**156**:76– 96.

[3] Lawless J. Statistical Models and Methods for Lifetime Data. Wiley Series in Probability and Statistics. New York, USA: Wiley; 2003.

[4] Lee E., Wang J. Statistical Methods for Survival Data Analysis. Wiley Series in Probability and Statistics. New York, USA: Wiley; 2003.

[5] Marchenko Y., Genton M. Multivariate log-skew-elliptical distributions with applications to precipitation data. Environmetrics. 2010;**21**:318– 340.

[6] Mitzenmacher M. A brief history of generative models for power law and lognormal distributions. Internet Mathematics. 2004;**1**(2):226– 251.

[7] Eeckhout J. Gibrat's law for (All) cities. American Economic Review. 2004;**94**:1429– 1451.

[8] Gabaix X. Power laws in economics: An introduction. Journal of Economic Perspectives. 2016;**30**:185– 206.

[9] Clarke R. Estimating trends in data from the Weibull and a generalized extreme value distribution. Water Resources Research. 2002;**38**:25-1– 25-10.

[10] Kleiber C., Kotz S. Statistical Size Distributions in Economics and Actuarial Sciences. New York: John Wiley & Sons; 2003.

[11] Asgharzadeh A., Nadarajah S., Sharafi F. Generalized inverse Lindley distribution with application to Danish fire insurance data. Communications in Statistics: Theory and Methods. Forthcoming.

[12] Ortega E.M.M., Lemonte A.J., Silva G.O., Cordeiro G.M. New flexible models generated by gamma random variables for lifetime modeling. Journal of Applied Statistics. 2015;**42** (10):2159– 2179.

[13] Aitchison J., Brown J.A.C. The Lognormal Distribution. Cambridge: University Press; 1957.

[14] McAlister D. The law of the geometric mean. Proceedings of the Royal Society of London. 1879;**29**:367– 376.

[15] Black F., Scholes M. The pricing of options and corporate liabilities. The Journal of Political Economy. 1973;**81**(3):637– 654.

[16] Clementi F., Gallegati M. Power law tails in the Italian personal income distribution. Physical A: Statistical Mechanics and Its Applications. 2005;**350**(2):427– 438.

[17] Burnecki K., Kukla G., Weron R. Property insurance loss distributions. Physical A: Statistical Mechanics and Its Applications. 2000;**287**(1):269– 278.

[18] Burnecki K., Misiorek A., Cizek P., Härdle W.K. and Weron R. Loss distributions. In: Statistical Tools for Finance and Insurance. Berlin, Heidelberg: Springer; 2005. pp. 289–317.

[19] Koch A.L. The logarithm in biology I. Mechanisms generating the log-normal distribution exactly. Journal of Theoretical Biology. 1966;**23**:276– 290.

[20] Limpert E., Stahel W.A., Abbt M. Log-normal distributions across the sciences: keys and clues. BioScience. 2001;**51**(5):341– 352.

[21] Tufféry S. Data mining et statistique décisionnelle. Paris, France: Technip; 2012.

[22] Pareto V. Cours d'économie politique. Geneva: Droz; 1896.

[23] Arnold B.C. Pareto Distributions. Fairland, Maryland, USA: International Co-operative Publishing House; 1983.

[24] Hill B.M. A simple general approach to inference about the tail of a distribution. Annals of Statistics. 1975;**3**:1163– 1174.

[25] Dominicy Y., Ilmonen P., Veredas D. Multivariate Hill Estimators. International Statistical Review. Forthcoming.

[26] Lomax K.S. Business failures: Another example of the analysis of failure data. Journal of the American Statistical Association. 1954;**49**:847– 852.

[27] Stoppa G. Proprieta campionarie di un nuovo modello Pareto generalizzato. Atti XXXV Riunione Scientica della Societa Italiana di Statistica, Padova: Cedam. 1990; pp. 137– 144.

[28] Benckert L.G., Sternberg I. An attempt to find an expression for the distribution of fire damage amount. Transactions of the 15th International Congress of Actuaries. 1957;**2**:288– 294.

[29] Andersson H. An analysis of the development of the fire losses in the north countries after the Second World War. Astin Bulletin. 1971;**6**:25– 30.

[30] Reed W.J., Jorgensen M. The double Pareto-lognormal distribution–A new parametric model for size distributions. Communications in Statistics: Theory and Methods. 2004;**33**(8):1733– 1753.

[31] Abdul-Moniem I.B., Abdel-Hameed H.F. On exponentiated Lomax distribution. International Journal of Mathematical Archive. 2012;**3**(5):2144– 2150.

[32] Shams T.M. The Kumaraswamy-generalized Lomax distribution. Middle-East Journal of Scientific Research. 2013;**17**(5):641– 646.

[33] Tahir M.H., Hussain M.A., Cordeiro G.M., Hamedani G.G., Mansoor M., Zubair M. The Gumbel-Lomax distribution: properties and applications. Journal of Statistical Theory and Applications. 2016;**15**(1):61– 79.

[34] Fisher R.A., Tippett L.H.C. Limiting forms of the frequency distribution of the largest or smallest member of a sample. Proceedings of the Cambridge Philosophical Society. 1928;**24**:180– 290.

[35] Embrechts P., Klü ppelberg C., Mikosch T. Modelling Extremal Events for Insurance and Finance. Berlin: Springer-Verlag; 1997.

[36] Guégan D., Hassani B.K. A mathematical resurgence of risk management: An extreme modeling of expert opinions. Frontiers in Finance and Economics. 2014;**11**(1):25– 45.

[37] Burke E.J., Perry R.H.J., Brown S.J. An extreme value analysis of UK drought and projections of change in the future. Journal of Hydrology. 2010;**388:**131– 143.

[38] Coles S. An Introduction to Statistical Modeling of Extreme Values. Berlin: Springer-Verlag; 2001.

[39] Finkenstädt B., Rootzén H. Extreme Values in Finance, Telecommunications and the Environment. London: Chapman & Hall/CRC; 2004.

[40] Jenkinson A.F. The frequency distribution of the annual maximum (or minimum) values of meteorological elements. Quarterly Journal of the Royal Meteorological Society. 1955;**81**:158– 171.

[41] Lindquist E.S. Strength of materials and the Weibull distribution. Probabilistic Engineering Mechanics. 1994;**9**(3):191– 194.

[42] Manwell J.F., McGowan J.G., Rogers A.L. Wind Energy Explained: Theory, Design and Application. New York, USA: Wiley; 2009.

[43] Sharif N., Islam N. The Weibull distribution as a general model for forecasting technological change. Technological Forecasting and Social Change. 1980;**18**(3):247– 256.

[44] Alvarado-Celestino E. Large forest fires: An analysis using extreme value theory and robust statistics [thesis]. University of Washington, USA; 1992.

[45] Klugman S.A., Panjer H.H., Willmot G. Loss Models: From Data to Decisions. New York: Wiley; 2008.

[46] Cooray K., Ananda M.M.A. Modeling actuarial data with composite lognormal-Pareto model. Scandinavian Actuarial Journal. 2005;**5**:321– 334.

[47] Scollnik D.P.M. On composite lognormal-Pareto models. Scandinavian Actuarial Journal. 2007;**1**:20– 33.

[48] Calderin-Ojeda E., Kwok C.F. Modeling claims data with composite Stoppa models. Scandinavian Actuarial Journal. Forthcoming.

[49] Teodorescu S., Vernic R. Some composite exponential-Pareto models for actuarial prediction. Romanian Journal of Economic Forecasting. 2009;**12**(4):82– 100.

[50] Teodorescu S., Vernic R. On composite Pareto models. Mathematical Reports. 2013;**15** (65):11– 29.

[51] Scollnik D.P.M., Sun C. Modeling with Weibull-Pareto models. North American Actuarial Journal. 2012;**16**(2):260– 272.

[52] Ciumara R. An actuarial model based on composite Weibull-Pareto distribution. Mathematical Reports. 2006;**8**(58):404– 414.

[53] Ramirez-Cobo P., Lillo R.E., Wilson S., Wiper M.P. Bayesian inference for double Pareto lognormal queues. The Annals of Applied Statistics. 2010;4(3):1533– 1557.

[54] Rubio F.J., Steel M. Bayesian modelling of skewness and kurtosis with two-piece scale and shape distributions. Electronic Journal of Statistics. 2015;**9**:1884– 1912.

[55] McDonald J., Xu Y. A generalization of the beta distribution with applications. Journal of Econometrics. 1995;**66**:133– 152.

[56] Sinner C., Dominicy Y., Ley C., Trufin J., Weber P. An interpolating family of size distributions. Forthcoming.

8

A Comparison Study on Performance of an Adaptive Filter with Other Estimation Methods for State Estimation in High-Dimensional System

Hong Son Hoang and Remy Baraille

Abstract

In this chapter, performance comparison between the adaptive filter (AF) and other estimation methods, especially with the variational method (VM), is given in the context of data assimilation problem in dynamical systems with (very) high dimension. The emphasis is put on the importance of innovation approach which is a basis for construction of the AF as well as the choice of a set of tuning parameters in the filter gain. It will be shown that the innovation representation for the initial dynamical system plays essential role in providing stability of the assimilation algorithms for stable and unstable system dynamics. Numerical experiments will be given to illustrate the performance of the AF.

Keywords: dynamical system, innovation process, filter stability, minimum mean square prediction error, simultaneous stochastic perturbation

1. Introduction

Consider the following data assimilation problem: Given the dynamical system

$$x_{k+1} = \phi(x_k) + w_k, \tag{1.1}$$

and the observations

$$z_{k+1} = H_{k+1}x_{k+1} + v_{k+1}, k = 0, 1, 2, \ldots, N \tag{1.2}$$

Here, $x_k \in R^n$ is the system state at k instant, $\phi(.) : R^n \to R^n$, $z_k \in R^p$ is observation vector, $H_k \in R^{p \times n}$ is the observation matrix, w_k, v_k are the model and observation uncorrelated noise sequences which are mutually uncorrelated and uncorrelated with x_0. The statistical characteristics of the entering random variables are given as

$$\begin{aligned} E[x_0] &= \overline{x}_0, E[x_0 x_0^T] = M_0, \\ E[w_k] &= 0, E[w_k w_l^T] = \delta_{kl} Q, \\ E[v_k] &= 0, E[v_k v_l^T] = \delta_{kl} R, E[w_k v_l] = 0, \\ E[(x_0 - \overline{x}_0) w_k^T] &= 0, E[(x_0 - \overline{x}_0) v_k^T] = 0. \end{aligned} \tag{1.3}$$

The problem we consider here is to estimate the system state x_k under the conditions that the dimension n of x_k is of order $10^6 - 10^8$, and there are uncertainties in statistics of the model and observational noises. Due to very large n, it is impossible to apply traditional estimation algorithms for producing the estimate $\hat{x}_k$ and that is the reason there exist different approximation algorithms for solving this estimation problem. Theoretically, the optimal in mean square error (MSE) estimate $\hat{x}_{k/N}$ based on the set of observations $Z[1, N] := \{z_1,\ldots, z_N\}$ is a filtered estimate for $N = k$ and smoothed estimate for $N > k$ [1, 2]. For the linear dynamical system Eqs. (**1.1**) and (**1.2**), the computation of $\hat{x}_k := \hat{x}_{k/k}$ can be efficiently performed using the Kalman filter (KF) [3] which is a sequential procedure. The KF provides also the equations for computation of estimation errors. If we are interested in obtaining $\hat{x}_{k/N}$—the best estimate for x_k based on $Z[1, N]$, the Kalman smoother can serve as an efficient algorithm for its computation. The KF approach, however, is inappropriate for solving estimation problems in high-dimensional systems. In this chapter, the high-dimensional system means the system whose state dimension is of order $10^6 - 10^8$. At the present and in the near future, the computer capacity, in both computational power and memory, is still very far to be sufficient to implement the KF in real time to produce the filtered estimate and to make corresponding forecast. For suboptimal schemes for atmospheric data assimilation based on the KF, see Ref. [4].

In this chapter, the emphasis is put principally on comparison of the AF with VM. For the review on the data assimilation methods in meteorology and oceanography, see Ref. [5]. To see more the advantages of the AF, we implement the extended KF (EKF) [6] in Section 6 and will compare its performance with that of the AF (the experiment with Lorenz system). The Cooper-Haines filter (CHF) [7], widely used in data assimilation in oceanography, is also applied in Section 7 to produce the estimate for the ocean state. It serves as a reference to be compared with that produced by the AF in high-dimensional setting.

In the next section, the variational method (VM), which is widely used in data assimilation for high-dimensional systems in meteorology and oceanography, is outlined. Section 3 provides the recently developed AF approach to data assimilation. The main idea of the AF is to take the innovation representation for the input-output system as a departure point to formulate the optimization problem, with the parameters of the filter gain as control variables. Section 4 presents the tools to implement the AF in a simple and efficient way which is adapted for high-dimensional setting. This includes the objective function, filter stability, structure of the error covariance matrix (ECM), gain parameterization, algorithm for optimization known as simultaneous perturbation stochastic approximation (SPSA). It is shown how the ECM is estimated using an ensemble of samples of prediction error (PE) and the hypothesis on separation of vertical and horizontal variables structure (SeVHS) in the ECM. Computational comparison between the VM and AF is also given here. Section 5 includes a simple numerical experiment showing in detail how work the VM and AF, the difficulties of the VM in searching optimal solution and why no similar difficulties are encountered in the AF. The more complicated experiment with chaotic system known as Lorenz system is done in Section 6. The difficulties encountered here concern extreme sensitivity of its solution to small errors in the initial condition. Section 7 presents the performance of different filters in the data assimilation experiment with the high-dimensional ocean model MICOM with the North Atlantic configuration. The conclusions are given in Section 8.

Notation: In the chapter, A^T denotes the transpose of the matrix A; $E[.]$, $E[.|.]$ denote the expectation and conditional expectation, respectively; $||A||_F$ denotes the Frobenius norm of a matrix A.

2. Variational method (VM)

Consider the problem of estimating $\{x_k\}$ in Eqs. (1.1)–(1.3). The VM consists of minimizing the following objective function

$$J[x_0, \ldots, x_N] = e_0 M_0^{-1} e_0 + \sum_{k=1}^{N} (z_k - H_k x_k)^T R^{-1} (z_k - H_k x_k), e_0 := x_0 - \overline{x}_0, \tag{2.1}$$

$$J[x_0, \ldots, x_N] \rightarrow \min{}_{[x_0, \ldots, x_N]} \tag{2.2}$$

$$\text{under the constraints (1.1)} \tag{2.3}$$

Thus, in the VM, we seek optimal solutions in the functional space (space of functions $\{x_k\}$). For systems of high dimension, this task is impossible to perform. The simplification is required. Suppose the system Eq. (1.1) is linear and perfect, that is,

$$x_{k+1} = \Phi_k x_k, k = 0, 1, \ldots \tag{2.4}$$

Expressing all x_k as functions of the initial state x_0,

$$x_k = \Phi(k, 0) x_0,$$
$$\Phi(k, l) = \Phi_{k-1} \ldots \Phi_l, (k > l), \Phi(k, k) = I, \tag{2.5}$$
$$I \text{ is the identity matrix of appropriate dimension}$$

and substituting $x_k, \forall k$ Eq. (2.5) into Eq. (1.1), at each kth observation instant, the following set of observations is available for x_0,

$$z_k^1 = H_k^1 x_0 + v_1^k, k = 1, 2, \ldots$$
$$H_k^1 := [(H_1\Phi(1,0))^T, \ldots, (H_k\Phi(k,0))^T]^T, \quad (2.6)$$
$$v_k^1 = [v_1^T, \ldots, v_k^T]^T.$$

Under the assumption on perfect model, the optimization problem Eqs. (2.1)–(2.3) is simplified as

$$J[x_0] \rightarrow \min_{[x_0]}, \quad (2.7)$$

$$J[x_0] := e_0^T M_0^{-1} e_0 + \sum_{k=1}^{N} (z_k - H'_k x_0)^T R_k^{-1} (z_k - H'_k x_0),$$
$$H'_k := H_k \Phi(k,0). \quad (2.8)$$

We have now the unconstrained optimization problem Eqs. (2.7) and (2.8) with the vector of unknown parameters $\theta := x_0$—the initial state. This problem can be solved using standard optimization techniques [8].

It is not hard to write out a solution to the problem Eqs. (2.7) and (2.8). For high-dimensional systems, there is no computational and memory resources to handle such implementation. In practice, the solution to the problem Eqs. (2.7) and (2.8) is found by solving iteratively the equation

$$\nabla_\theta J[\theta] = 0,$$
$$\nabla_\theta J[\theta] := [\partial J/\partial \theta_1, \ldots, \partial J/\partial \theta_1]^T. \quad (2.9)$$

Comment 2.1. Usually, finding a solution to Eqs. (**2.7**) and (**2.8**) is a heavy task: in addition to storing the model solution produced by the direct model, minimization requires 20–30 iterations to reach a relatively good approximate solution.

Comment 2.2. Writing out $\nabla_\theta J[\theta]$ shows that solving Eq. (**2.9**) requires

$$(H'_k)^T y = \Phi_k^T \Phi_{k-1}^T \ldots \Phi_1^T H_k^T y \quad (2.10)$$

for some y. As Φ_k^T is impossible to store, the approach known as adjoint equation (AE) is used which requires to construct AE code for computing the product $\Phi_k^T y$. Each iteration in minimization of Eqs. (2.7) and (2.8) thus requires one integration of the model over the assimilation period, followed by one adjoint integration. The cost of one adjoint integration is about twice the cost of one direct integration, so that one minimization requires the equivalent of between 50–100 integrations of the model over the assimilation period (p. 205 [9]). In the next section, we see that the SPSA can also be used to solve this problem at a much lower cost.

Comment 2.3 As θ—initial state—has a physical meaning, it is important to introduce constraints on the appropriate physically realistic structure of the correction for θ during the estimation process. A poor (in the physical sense) structure of the guess for the initial state can lead to large estimation errors.

3. Adaptive filtering (AF)

To overcome the difficulties listed in Comments 2.1–2.3, an adaptive filtering (AF) has been proposed in [10]. The main difference of the AF with the VM is lying in the choice of innovation representation for the original input-output system Eqs. (1.1) and (1.2) as a departure point to formulate the optimization problems. It is well known that under standard conditions, the optimal in MSE estimate $\hat{x}_k$ can be obtained by the KF. As the innovation process for the system output in the KF forms a white sequence, Kailath [11] has developed an innovation approach, in an elegant way, to derive the optimal filter for more general linear systems like nonstationary, filtering problems with Markovian processes for the model and observation errors. The innovation approach to linear least-squares approximation problems is first to "whiten" the output data and then to treat the resulting simpler white-noise observation problem. Consider the observation (output) sequence z_k. The innovation process, associated with z_k, is written as

$$\zeta_k = z_k - E\left[z_k | z_{k-1}^1\right], \tag{3.1}$$

Under standard conditions (Gaussianness, uncorrelated noise sequences ...), $E\left[z_k|z_{k-1}^1\right] = H_k\hat{x}_{k/k-1}$ hence

$$\zeta_k = z_k - H_k\hat{x}_{k/k-1}, \hat{x}_{k/k-1} = \Phi_{k-1}\hat{x}_{k-1}, \tag{3.2}$$

where $\hat{x}_{k/k-1}$ is an optimal in MSE one-step ahead prediction for x_k given z_{k-1}^1. Using ζ_k instead of z_k, one can write out the formula for the estimate $\hat{x}_k$ and the KF under standard conditions. The filter has the form

$$\begin{aligned} \hat{x}_k &= \Phi_k\hat{x}_{k-1} + K_k\zeta_k, \\ K_k &= M_kH_k^T\left[H_kM_kH_k^T + R_k\right]^{-1} \end{aligned} \tag{3.3}$$

where M_k is the ECM for the prediction $\hat{x}_{k/k-1}$. This matrix is found as a solution to the Riccati equation

$$M_k = \Phi_kP_k\Phi_k^T + Q_k, P_k = \left[I - K_kH_k\right]M_k. \tag{3.4}$$

Due to the very expensive computational burden in time stepping the ECM M_k in Eq. (**3.4**) as well as insufficient memory storage, the KF is impractical for solving data assimilation problems in very high-dimensional setting. The idea of the AF is based on the fact that when the filter is optimal, the innovation ζ_k has a minimum variance. If we assume that the gain K_k belongs to a set of parameterized gains, that is,

$$K_k = K_k(\theta), \theta \in \Theta, \tag{3.5}$$

the optimal AF can be considered as that in some class of parameterized filters of a given structure. The following objective function is introduced

$$J(\theta) = E[\Psi(\theta)] \rightarrow \min_{\theta \in \Theta}, \Psi(\zeta_k) = ||\zeta_k||^2_{\Sigma_k^{-1}}, ||\zeta_k||^2_{\Sigma_k^{-1}} := < \zeta_k, \Sigma_k^{-1}\zeta_k > . \quad (3.6)$$

In Ref. [12], the different classes of parameterized filters are found which belong to the class of stable reduced-order filters (ROF) [10, 13].

As an example for one class of ROFs, consider

$$K_k = P_{r,k}K_{e,k} \quad (3.7)$$

where $K_{e,k} : R^p \rightarrow R^{n_e}$ represents the gain, mapping the innovation vector from the observational space to the reduced space R^{n_e} of dimension $n_e \leq n$; $P_{r,k}$ is mapping the reduced space R^{n_e} to the full space R^n. The choice of a reduced space is of primary importance since it depends on the main characteristics of the filter known as stability. As proved in [12], under detectability condition, stability of the filter is ensured by forming the columns of $P_{r,k}$ from unstable and stable eigenvectors (or singular vectors, Schur vectors) of the fundamental matrix Φ_k, and one can choose

$$K_{e,k} = H_{e,k}^T[H_{e,k}H_{e,k}^T(k) + R_k]^{-1}, H_{e,k} := H_kP_{r,k}, \quad (3.8)$$

One class of parameterized filters is (Section 5.2.2 in Ref. [12])

$$\begin{gathered} K_k(\theta) = P_{r,k}\Lambda K_{e,k}(\theta), \\ \Lambda = \text{diag } [\theta_1,\ldots, \theta_{n_e}], \\ 1\text{–}1/|\phi_i| < \grave{o}_1(i) \leq \theta_i \leq \grave{o}_2(i) < 1 + 1/|\phi_i|. \end{gathered} \quad (3.9)$$

if ϕ_i is an unstable or neutral eigenvector of Φ. For the stable ϕ_i, we have

$$0 < \grave{o}_1(i) \leq \theta_i \leq \grave{o}_2(i) < 2, \quad (3.10)$$

4. Differences between VM and AF

4.1. Batch data formulation

We list now the main differences between two approaches VM and AF from which it becomes clear what are the advantages of the AF over the VM.

To make easier comparison between two approaches, let us write out the objective function Eq. (**3.6**) using a representation in a sample space

$$\begin{gathered} J_N(\theta) \rightarrow \min_{\theta \in \Theta}, \\ J(\theta) \approx J_N(\theta) = \frac{1}{N}\Sigma_{k=1}^N(z_k - H\Phi_k\hat{x}_{k-1}(\theta))^T\Sigma_k^{-1}(z_k - H\Phi_k\hat{x}_{k-1}(\theta)) \end{gathered} \quad (4.1)$$

Mention that in practical implementation of the AF, the optimization algorithm is not constructed on the basis of Eq. (4.1), but on Eq. (3.6). That is, due to the fact that Eq. (**4.1**) is written in a batch form which requires to make optimization over the time interval $[1, N]$

resulting in a very high computational burden. Minimizing Eq. (**3.6**) allows to apply SPSA method which is much less consuming for both computational and memory requirements. Below the main differences between VM and AF are listed:

(D1) Dynamical system (DS): if in Eq. (**2.1**), the DS is the initial system Eq. (**1.1**); in Eq. (**4.1**), the DS is the filtering Eq. (**3.3**). This difference has an interesting consequence: if in practice, there is very little known about statistics of w_k, the sequence ζ_k is observed, and hence, it is possible to estimate the statistics of ζ_k.

(D2) The system noise w_k in Eq. (**1.1**) is white, while in Eq. (**3.3**), ζ_k is a white sequence only if the filter is optimal. That allows us to easily apply different statistical tests for verifying the optimality of the assimilation procedure.

(D3) Control variable x_0 in the VM is the initial state, whereas the control variable in Eq. (**3.6**) is the parameter vector θ.

This difference has an important consequence: as x_0 has to be of precise physical meaning (depending, for example, on the ocean domain of interest), the structure for the guess $\theta_0 := \hat{x}_0^0$ (for the initial state) as well as correction $\delta\hat{x}_0^\nu$, generated by iterative algorithm, must be chosen carefully so that at each ν iteration, the estimate $\hat{x}_0^\nu$, $\hat{x}_0^\nu = \hat{x}_0^{\nu-1} + \delta\hat{x}_0^\nu$, must be of physically realistic structure. This is not an easy task. On the other hand, in the AF, the parameters usually are immaterial [see θ Eq. in (**3.10**)]; hence, the choice of structure for θ is of no importance.

(D4) Suppose the DS Eq. (**1.1**) is unstable. It implies that the error in estimating x_0 will grow during integration of the direct and AE. As for the AF, by its construction, the filtering system Eq. (**3.3**) remains stable. This can be seen by representing the filtering Eq. (**3.3**) through its fundamental matrix L_k,

$$\begin{aligned} \hat{x}_k &= L_k\hat{x}_{k-1} + K_k v_k \\ L_k &= (I - K_k H_k)\Phi_k. \end{aligned} \tag{4.2}$$

As shown in Ref. [12], the filter Eq. (**4.2**) is stable under the conditions Eqs. (**3.9**) and (**3.10**). It means that the filtering error is bounded during model integration since the parameters θ_i are lying in the interval guaranteeing a stability of the filter Eq. (**4.2**).

(D5) Return to the objective function Eqs. (**2.1**)–(**2.3**). First taking the derivative of the objective function Eqs. (**2.1**)–(**2.3**) wrt (with respective to) x_0, we have

$$\begin{aligned} \frac{1}{2}\nabla_{x_0} J[\hat{x}_0^\nu] &= M_0^{-1}e_0^\nu - \sum_{k=1}^{N}\Phi^T(k,0)H_k^T R_k^{-1}(z_k - H_k\Phi(k,0)x_0^\nu) \\ &= M_0^{-1}e_0^\nu - \sum_{k=1}^{N}\Phi^T(k,0)H_k^T R_k^{-1}(H_k\Phi(k,0)e_0^\nu + v_k), e_0^\nu := x_0 - \hat{x}_0^\nu. \end{aligned} \tag{4.3}$$

One sees that for a batch of N observations, Eq. (**4.3**) requires computation of N terms (without counting for the term $M_0^{-1}e_0^\nu$). The kth term is associated with the assimilation instant k, and one needs to compute first $\mu_k := \Phi(k,0)e_0^\nu$, that is, to integrate k times the direct model Φ_κ for $\kappa = 1,\ldots, k$ and next to integrate backward (k times also) the AE Φ_κ^T from $\kappa = k$ to $\kappa = 1$, that is, to compute $\Phi^T(k,0)H_\kappa^T R_\kappa^{-1}(H_k\mu_k + v_k)$. The larger the k, the bigger the amplification of the initial error e_0^v and the observation error v_k. The error e_0^v is amplified doubly since it is integrated by the direct and adjoint models. But the amplification of v_k (and w_k when $w_k \neq 0$) is most worrying since it is integrated in the gradient estimate, making the gradient direction to be, possibly, completely erroneous.

4.2. Implementation of AF

4.2.1. The choice of criteria Eq. (3.6)

The choice of Eq. (3.6) is important in many aspects in order to obtain a simple and efficient data assimilation algorithm. The idea lying in the criteria Eq. (3.6) is to select some pertinent parameters as a control variables for minimizing the mean of the cost function $\Psi(\zeta_k)$. For example, for the class of filters Eqs. (3.3), (3.7)–(3.10) the vector $\theta = (\theta_1, \ldots, \theta_{n_e})^T$ can be chosen as control vector for the problem Eq. (3.6).

The solution to the problem Eq. (3.6) can be found iteratively using a stochastic optimization (SA) algorithm

$$\theta_{k+1} = \theta_k - a_k \nabla_\theta \Psi(\zeta_{k+1}) \tag{4.4}$$

where $\{a_k\}$ is a sequence of positive scalars satisfying some conditions to guarantee a convergence of the estimation procedure. The standard conditions are

$$a_k \to 0, \sum_{k=1}^{\infty} a_k = \infty, \sum_{k=1}^{\infty} a_k^2 < \infty \tag{4.5}$$

The algorithm Eq. (4.4) is much more simple [compared to the computation of Eq. (4.3)] since it requires, at the kth assimilation instant, to compute only the gradient of the sample cost function $\Psi(\zeta_k)$. The gradient $\nabla_\theta \Psi(\theta_k)$ of the sample objective function $\Psi(\theta_k)$ can be computed using the AE approach (in what follows, for simplicity, the subscript k will be omitted to shorten the notations).

$$\frac{1}{2}[\delta\Psi(\zeta_{k+1})]_{\theta_k} = -(H\Phi P_r \delta\Lambda K_e \zeta_k, \zeta_{k+1}) = -(\delta\Lambda K_e \zeta_k, \zeta'_{k+1}), \zeta'_{k+1} := P_r^T \Phi^T H^T \zeta_{k+1} \tag{4.6}$$

Thus, minimization of Eq. (3.6) by gradient-based SA algorithm requires only one integration of the direct model and one backward integration of the AE code: direct integration of $\hat{x}_k$ for producing the forecast $\hat{x}_{k+1/k} = \Phi\hat{x}_k$ and backward integration $\Phi^T H_k \zeta_{k-1}$ in computation of ζ'_{k+1}. For the structure of the gain Eq. (3.9), the objective function Ψ is quadratic wrt θ; hence, one can find easily the optimal parameters.

A less computational burden can be achieved by measuring the sample objective function (but not based on a gradient formula): instead of computing the gradient by Eq. (4.6) based on AE, one can approximate the gradient using the values of the cost function [on the basis of finite difference scheme (FDSA)]. Traditionally, the ith component of the gradient can be approximated by

$$\nabla_{\theta_i}\Psi(\theta_k) = g_i = [\Psi(\theta_k + c_k e_i) - \Psi(\theta_k - c_k e_i)]/(2c_k) \tag{4.7}$$

where e_i is the unit vector with 1 in the ith component, 0 otherwise.

It is seen that FDSA algorithms do not require the formula for the gradient. However, for the high-dimensional systems ($n \approx O(10^6)$–$O(10^7)$), this algorithm is inapplicable due to component-wise derivative approximation: for approximation of each partial derivative of the cost function, we need to make two integrations of the direct model.

In order to overcome the difficulties with very high dimension of θ, recently the class of algorithms known as simultaneous perturbation SA (SPSA) receives a great interest [14, 15]. The algorithm SPSA is of the same structure as that of FDSA Eq. (**4.7**), with the difference residing in the way to perturb stochastically and simultaneously all the components of θ. Concretely, let $\Delta_k = (\Delta_{k,1},\ldots, \Delta_{k,n})^T$ be a random vector, $\Delta_{k,i}, i = 1,\ldots, n$ are Bernoulli independent identically distributed (iid). The gradient of the objective function is estimated as

$$\begin{gathered} \nabla_\theta \Psi(\theta_k) = g = (g_1,\ldots, g_n)^T, \\ g = [\Psi(\theta_k + c_k\Delta_k) - \Psi(\theta_k - c_k\Delta_k)]\Delta_k^{-1}/(2c_k), \\ \Delta_k = (\Delta_{k,1},\ldots, \Delta_{k,n})^T, \Delta_k^{-1} := (1/\Delta_{k,1},\ldots, 1/\Delta_{k,n})^T. \end{gathered} \tag{4.8}$$

It is seen that in the SPSA, all the directions are perturbed at the same time (the numerator is identical in all n components). Thus, SPSA uses only two (or three) times integrations of the model, independently on the dimension of θ which makes it possible to apply to high-dimensional optimization problems. Generally, SPSA converges in the same number of iterations as FDSA, and it follows approximately the steepest descent direction, behaving like the gradient method [14]. On the other hand, SPSA, with the random search direction, does not follow exactly the gradient path. On average, though, it tracks the gradient nearly because the gradient approximation is an almost unbiased estimator of the gradient, as shown in Ref. [15].

For the SPSA algorithm, the conditions for $\{a_k\}$ and $\{c_k\}$ are

$$\begin{gathered} a_k > 0, c_k > 0, \quad a_k \to 0, \quad c_k \to 0, \\ \sum_{k=1}^{\infty} a_k = \infty, \sum_{k=1}^{\infty} (a_k/c_k)^2 < \infty \end{gathered} \tag{4.9}$$

4.2.2. *On the operator P_r*

As shown in Ref. [12], *span*$[P_r]$–the subspace, spanned by the columns of P_r, must be chosen so that the filter gain K ensures a stability of the filter. Mention that even the KF may suffer from instability. To ensure filter stability, P_r is constructed from all unstable and neutral eigenvectors of the fundamental matrix Φ (or real Schur vectors (ScVs), singular vectors). In practice, we choose P_r to be consisting of the column vectors of S (called *S-PE samples*)

$$S = \Phi X \tag{4.10}$$

which are results of integration of leading ScVs (columns of X). The columns in S have the meaning of the PE for the system state and are used to approximate the ECM M. As to the

ScVs, they are preferred to eigenvectors (or singular vectors) because the ScVs are real, and their computation is numerically stable. Mention that computation of singular vector requires also adjoint code. The ensemble of columns of S plays the same role as an ensemble PE samples in the ensemble-based filtering technique for approximating the background ECM [16, 17].

4.2.3. *On separation of vertical and horizontal variables structure in ECM [18]*

Let us consider the situation when the DS is described by PDEs. The state vector at the time instant k is $x_k = x_k(i,j,l)$ where (i,j,l) represents a grid point in three dimensional space. Introduce the stabilizing structure for the filter gain [12]

$$\begin{aligned} K_k &= M_k H_k^T [H_k M_k H_k^T + R_k]^{-1}, \\ M &= M_d, M_d := P_r \Lambda P_r^T, \end{aligned} \tag{4.11}$$

where Λ is symmetric positive definitive. As shown in Ref. [12], one can choose Λ to be diagonal with diagonal elements serving to regularize the amplitude of ECM. The matrix M in Eq. (**4.11**) is ECM, and if it is computed on the basis of the Riccati equation (**3.6**), K_k is known as the KF gain. For $M = M_d$, computation of M is realizable if the reduced dimension n_e is not too large. Actually the number n_e of the ensemble size is of order $O(100)$ which is too small for M to be a good approximation for the true ECM. In Ref. [18], it is assumed that the estimated ECM is a member of the class of ECMs with separation of vertical and horizontal variables structure (SeVHS). Mention that this hypothesis is not new and used in modeling the ECM in meteorological data assimilation [19]. The optimal ECM is found as a solution of the minimization problem

$$\begin{aligned} &J(\theta) = E||M_d - M_v(\theta_1) \otimes M_h(\theta_2)||_F^2, \\ &||.||_F \text{ denotes matrix Frobenious norm }, \\ &J(\theta) \rightarrow \min{}_{\theta}, \theta = (\theta_1^T, \theta_2^T)^T. \end{aligned} \tag{4.12}$$

As the number of vertical layers in the today's numerical models is of order $O(10)$, all elements of the vertical ECM M_v (included in θ_1) can be considered as tuning to be estimated. As to M_h, it is often chosen in analytical form (e.g., the 1st or 2nd order autoregressive models). The parameters like correlation length can be selected as components of the control vector θ_2 in M_h.

Using dominant real Schur vectors has advantages that they are real and their computation is stable [12] while the computation of eigenvectors is unstable, and they may be complex.

4.3. Computational comparison between VM and AF

We give a brief comparison (computational burden) between the VM and AF algorithms. **Table 1** shows the number of elementary arithmetic operations required for implementation of VM and AF filtering algorithms based on AE tool. A smaller number of operations are required for the AF if the SPSA method is used (no need of Φ^T). Here, for simplicity, n^2

operations are accounted for the product $\Phi^T y$ (the same number as that required for Φy). In the AF, we assume that the full ECM M is used, whereas in the AROF (adaptive ROF), $M := P_r P_r^T$ as shown in Eq. (**4.6**). These numbers are calculated on the basis of Eqs. (**2.1**) and (**2.3**) since they represent the most computational burdens for these two algorithms. These numbers are rounded up to the dimensions of the entry matrices. Here, N_{it_o} is a number of iterations required to solve the minimization problem (**2.1**)–(**2.3**); N_{it} is a number of iterations required to solve the equation

$$\Xi y = \zeta_k, \Xi := [HMH + R]. \tag{4.13}$$

In the VM algorithm, the computation of M_0^{-1}, R^{-1} is not taken into account. In **Table 1**, there is also the number of operations required for the AROF, when the ECM M is given in the product decomposition form $M = P_r P_r^T$, $P_r \in R^{n \times n_e}$. In this situation, instead of n^2 operations in the AF, we need to perform $2nn_e$ operations. For $n_e << n$, much less computational and memory requirements are needed to perform the AROF.

VM	$(n^2(N^2 + N) + N(2np + p^2) + n^2)N_{it_o}$
AF	$((n^2 + 2np + p^2)N_{it} + 2(np + n^2))N$
AROF	$((2nn_e + 2np + p^2)N_{it} + np + 2nn_e)N$

Table 1. Number of elementary arithmetic operations.

To have the idea on how work in practice the VM and AF, here the examples of experiments with two numerical models MICOM (see Section 7) and HYCOM (Hybrid Coordinate Ocean Model) [20] developed at SHOM, Toulouse, France. The first experiment is performed with the MICOM model. The observations are available each 10 days (*ds*) during 2 years. For the MICOM (state dimension $n = 3 \times 10^5$), the 10 *ds* forecast requires 45 s (supercomputer Caparmor, IFREMER, France, sequential run). The 2-year integration takes 54.45 min (73 × 45 s). The AF needs 54.45 min × 3 = 164 min (or 2 h 45 min) to perform the assimilation experiment for the 2-year period. In this context, the VM requires between 5.7 ds and 11.4 ds to perform the experiment (hypothesis 50–100 times integration of the MICOM over the 2-year window, Section 2.2). As to the HYCOM (state dimension $n = 7 \times 10^7$), the observations are available each 5 ds. The 5 ds forecast requires 1 h (supercomputer Beaufix, Météo, France, parallel run, 62 processors). Two year integration requires 146 × 1 h = 146 h (or 6 ds). The AF needs, hence 18 ds to make the 2-year experiment. As to the VM, the experiment requires between 304 ds and 608 ds. That is one of the reasons why in operational setting one has to choose a short window for assimilating the observations by the VM.

Comment 4.1. Looking at **Table 1**, one sees that the dominant numbers n_d of operations in the VM and AF are $n_d(VM) = n^2 N^2 N_{it_o}$ and $n_d(NAF) = n^2 N N_{it}$. If we assume that $N_{it_o} \approx N_{it}$ (in fact, the number of iterations for solving the optimization problem Eqs. (**2.7**) and (**2.8**) is often larger than the number of iterations for solving the system of equations (**4.13**); it is more critical for the VM when the DS is nonlinear); the number $n_d(VM)$ is N times larger than $n_d(NAF)$.

5. Simple numerical examples: scalar case

5.1. Estimation problem

Consider the simple scalar dynamical process $x(t) = sin(t)$. As $\dot{x}(t) = cos(t)$, for $t_k = k\delta t$, using the approximation $x(t_k + \delta t) - x(t_k) = cos(t_k)\delta t$, one has the following discrete dynamical system

$$
\begin{aligned}
x_{k+1} = \phi_k x_k + u_k + w_k, \phi_k = \phi = 1, u_k = cos(k\delta t),\\
E(w_k) = 0, E(w_k v_l) = \sigma_w^2 \delta_{kl}, k = 0, 1, \ldots, N-1.
\end{aligned}
\tag{5.1}
$$

In Eq. (**5.1**), w_k represents the Gaussian model error. Suppose at each t_k moment, we observe the state x_k corrupted by the Gaussian noise v_k, hence

$$
\begin{aligned}
z_k = hx_k + v_k, h = 1, k = 1, \ldots, N,\\
E(v_k) = 0, E(v_k v_l) = \sigma_v^2 \delta_{kl}, k \neq l
\end{aligned}
\tag{5.2}
$$

where δ_{kl} is the Kronecker symbol. Suppose that the true initial state $x_0^* = 0.5$. The problem we study here is to estimate the system state x_k based on the set of observations $z_k, k = 1, \ldots, N$.

5.1.1. Experiment: cost functions

In the experiment, $\delta t = 0.01$, $N = 1000$. The two methods, VM and AF, will be implemented to produce the system estimates. We study two situations: (S1) the model is considered as perfect, that is, $w_k = 0$; and (S2) there exists the model error w_k with the variance $\sigma_w^2 = 0.001$. As to v_k, $\sigma_v^2 = 0.1$ in both cases.

Let $w_k = 0$. To see the advantages of the AF over VM in finding optimal solutions, **Figures 1** and **2** display the curves (time averaged variance of distances between the true trajectory and those resulting from varying the control variables) as functions of tuning parameters in these two methods: the initial state $\theta := x_0$ and the parameter $\theta := \lambda$ (see Eq. (**3.9**)). We remark that for the filtering system (**5.1**), (**5.2**), as $\phi = 1$, the system has one stable eigenvalue and one stable eigenvector (singular value and Schur decompositions are of the same structure). The filter fundamental matrix $L = (1-Kh)\phi = (1-K)$ is stable if $K \in (0, 2)$. For the gain structure (**4.11**), L is stable for any $M > 0$ since $K = \frac{M}{M+\sigma_r^2}$ for $h = 1$ and $\sigma_v^2 = 0.1 > 0$. We have then $K \in (0, 1) \subset (0, 2)$. Thus, one can choose $\theta := M > 0$ as tuning parameter. This structure is of less interest compared to Eq. (**3.9**) since θ enters in K in a nonlinear way, and in fact, K is allowed to vary only in the interval $(0, 1)$. For Eq. (**3.9**), $P_r = 1, H_e = 1$ hence $K_e = \frac{1}{1+\sigma_r^2} < 1$ and the filter is stable if θ satisfies Eq. (**3.10**). For this structure, the filter is stable for $K \in (0, 2)$. We will select the last structure as a departure point to optimize the AF performance.

From **Figure 1**, it is seen that the curve "noise-free" is equal to 0 when $\hat{x}_0 = 0.5$ for S1, but the "noisy" attains the minimal value 0.121 at $\hat{x}_0 = 0.6$. We note that almost the same picture is obtained for the cost function (**2.1**) (time averaged variance of the distance between $\hat{x}_k, k = 1, \ldots, 1000$ and observations $z_k, k = 1, \ldots, 1000$) subject to $\hat{x}_0 \in [-1 : 1]$. Despite the fact that the curves in **Figure 1** are quadratic, it is impossible to find the true initial state in the noisy situation since almost the same curves are obtained for the cost function Eq. (**2.1**).

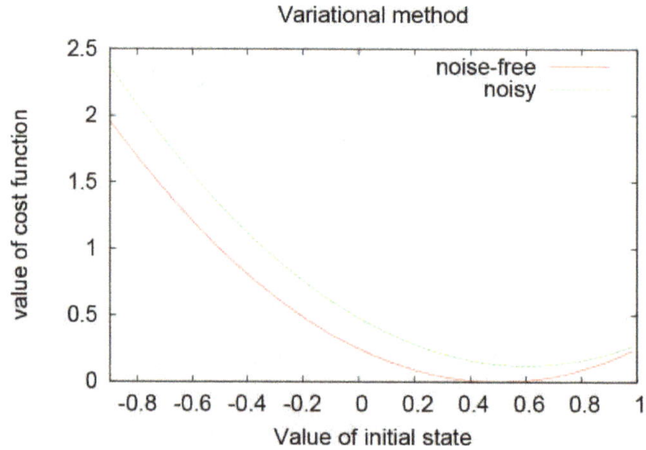

Figure 1. VM: cost functions resulting from perfect model and that with a model error.

Figure 2 presents the same curves as those in **Figure 1** resulting from application of the filter by letting the parameter θ in the gain vary in $\theta \in (0:2)$. **Figure 2** shows that for the both curves "noise-free" and "noisy", the minimal values are attained at $\theta = 1.1$ for both situations S1, S2. Moreover, the two minimal values are identical. The same picture is observed for the cost function Eq. (**4.1**) as function of $\theta \in (0:2)$. It means that independently on whether the model is perfect or not, AF formulation allows optimization algorithms to find the optimal value for θ, hence, to ensure optimality of the filter.

Two curves "noisy" in **Figures 1** and **2** show that when the model is noisy, the minimal value of the curve "noisy" (VM) in **Figure 1** is much higher (it is equal to 0.121) than that in **Figure 2** (0.009, for filtering). This fact is in favour of the choice of a short window for assimilating observations by the VM.

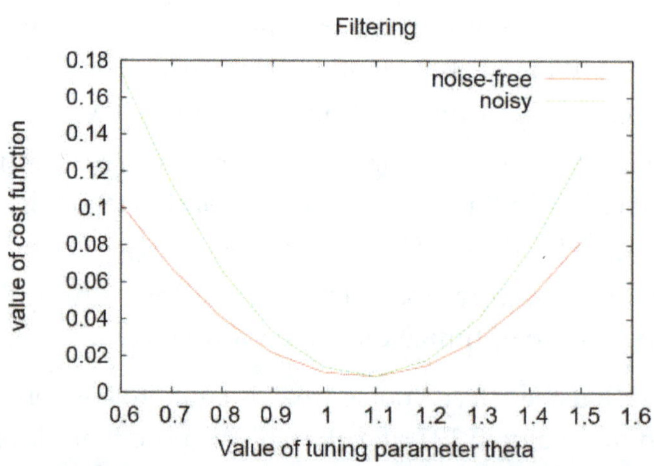

Figure 2. Filtering: cost functions resulting from perfect model and that with model error.

6. Numerical experiment: Lorenz system

6.1. Lorenz equations

The Lorenz attractor is a chaotic map, noted for its butterfly shape. The map shows how the state of a dynamical system evolves over time in a complex, non-repeating pattern [21].

The attractor itself and the equations from which it is derived were introduced by Edward Lorenz [21], who derived it from the simplified equations of convection rolls arising in the equations of the atmosphere.

The equations that govern the Lorenz attractor are:

$$\frac{dy_1}{dt} = -\sigma(y_1 - y_2), \frac{dy_2}{dt} = \rho y_1 - y_2 - y_1 y_3, \frac{dy_3}{dt} = y_1 y_2 - \beta y_3, \tag{6.1}$$

where σ is called the Prandtl number, and ρ is called the Rayleigh number. All σ, β, $\rho > 0$, but usually $\sigma = 10$, $\beta = 8/3$ and ρ is varied. The system exhibits chaotic behavior for $\rho = 28$ but displays knotted periodic orbits for other values of ρ.

6.2. Numerical model

In the experiments, the parameters σ, ρ, β are chosen to have the values 10, 28, and 8/3 for which the "butterfly" attractor exists.

The numerical model is obtained by applying the Euler method (first-order accurate method) to approximate Eq. (**6.1**). Symbolically, we have

$$y(t_{k+1}) = F'(y(t_k)), y(t_k) := (y_1(t_k), y_2(t_k), y_3(t_k))^T, \tag{6.2}$$

where $\delta t := t_{k+1} - t_k$ is the *model* time step. The observations arrive at the moments T_k and $\Delta T_k := T_{k+1} - T_k$. The experiment setup is similar to that described in Ref. [22].

6.3. Observations: assimilation

The corresponding $\delta t = 0.005$, $\Delta T_k = 1$, hence the sequence of observations is given by $z(k) := z(T_k), k = 1,\ldots, N_o$. The dynamical system corresponding to the transition of the states between two time instants T_k and T_{k+1} is denoted as

$$x_{k+1} = F(x_k) + w_k \tag{6.3}$$

In Eq. (**6.3**), w_k simulates the model error. The sequence w_k is assumed to be a white noise having variance 2, 12.13 and 12.13 respectively. The observation system is then given by

$$z_k = Hx_k + v_k \tag{6.4}$$

where the operator $H = [h_1^T, h_2^T]^T$, $h_1 = (1, 0, 0), h_2 = (0, 0, 1)$, that is, the first and third components x_1, x_3 are observed at each time instant $k = 1, \ldots, 100$. The noise sequence v_k is white with zero mean and variance $R = 2I_2$ where I_n is the unit matrix of dimension n. The initial estimate in all filters is given by the initial condition $\hat{x}(0) = (1,-1,24)^T$.

The true system state x^* is modeled as the solution of Eq. (**6.3**) subject to $x_0^* = (1.508870,-1.531271,25.46091)^T$.

The problem considered in this experiment is to apply the extended KF (EKF), nonadaptive filter (NAF), and adaptive filter (AF) to estimate the true system state using the observations z_k, $k = 1,2,\ N_o$ and to compare their performances.

Here, the NAF is in fact the prediction error filter (PEF). Mention that the PEF is developed in Ref. [23] in which the prediction error ECM is estimated on the basis of an ensemble of PE samples, that is,

$$M = \frac{1}{T-1}\sum_{k=1}^{T} B_s(k), B_s(k) = \sum_{k=1}^{L} \delta x_k^{(l)} \delta x_k^{(l),T}, \tag{6.5}$$

where $\delta x_k^{(l)}, l = 1,\ldots,\ L$ are members of the set of L S-PE samples obtained by $L+1$ integrations of the model from the reference state and L perturbed states which grow in the directions of the L dominant Schur vectors.

The filter gain is taken in the following form

$$K = MH^T\Sigma^{-1}, \Sigma = HMH^T + R \tag{6.6}$$

which is time invariant. At the same time, for the comparison purpose, the EKF is also used for assimilating the observations.

Figure 3 shows the evolution of the prediction errors resulting from three filters: NAF, AF, and EKF. It is of no surprise that the NAF has produced the estimates with larger estimation error. By adaptation, however, it is possible to obtain the AF, which improves significantly the PEF and even behaves better than the EKF.

Mention that the VM is much less appropriate for assimilating the observations in the Lorenz model due to the choice of the initial state as control vector. For simplicity, we simulate the situation when all three components of the system state are observed in additive noise, i.e. with $H = I_3$. **Figure 4** displays time averaged variances of the difference between the true trajectory and model trajectory (denoted as $AV(x^*,\hat{x})$), resulting from varying the third component of the initial state for two situations of noise-free and noisy models. Namely, we initialize the model by the initial state, which is the same as the true one $x_0^* = (1.508870,-1.531271,25.46091)^T$, with the difference, that the third component $\hat{x}_3(0)$ is varying in the interval $[24.5 : 26.5]$. The global minimum is attained at $x_0^*(3) = 25.46091$ as expected. However, if the system is initialized by the estimate in a vicinity, even not so far from $x_3^*(0)$, there is no guarantee that the VM can approach the true initial condition. For the noisy model, the global minimum is not attained at $x_0^*(3)$. As for the PEF, the function $AV(x^*,\hat{x})$ is quadratic wrt to the gain parameter, for both situations of noise-free or noisy models, as seen in **Figure 5**: here, the sample cost function (**4.1**) is computed over all assimilation period, by varying the third parameter θ_3 in the gain (related to the third observed component of the system state).

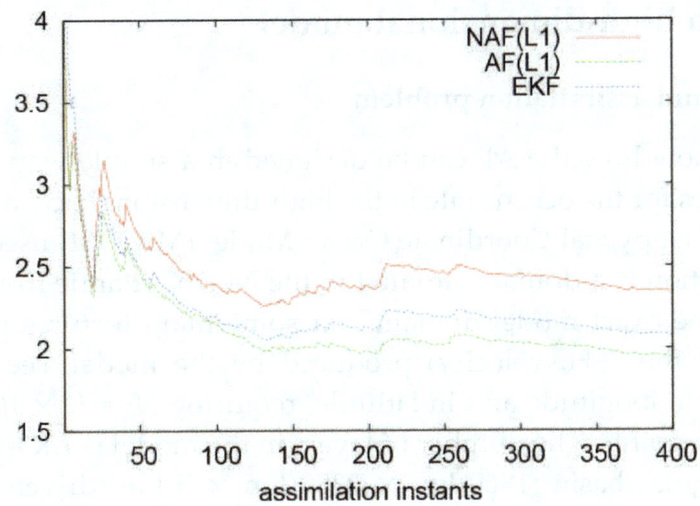

Figure 3. Prediction errors resulting from three filters: nonadaptive filter (NAF), adaptive filter (AF), and EKF.

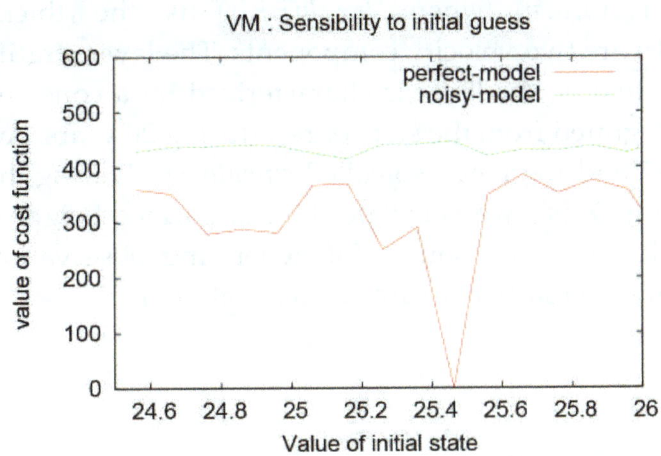

Figure 4. Time averaged variance between the true trajectory and model trajectory in the VM as a function of perturbed third component of the initial state. The global minimum is attained at the true initial condition, but there is no guarantee for the VM to approach the true initial state (no-noisy model). For noisy model, the global minimum is not attained at the true initial state. The curve "noisy model" is scaled by the factor $C = 1/15$.

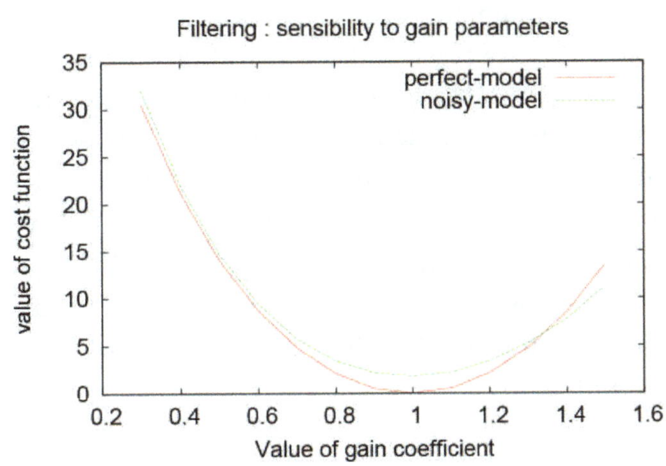

Figure 5. Cost function in the PEF as a function of perturbed third gain parameter θ_3. It is seen that in the PEF, the cost function is quadratic wrt to the gain parameter in both situations of noise-free and noisy models. The curve "noisy model" is scaled by the factor $C = 1/50$.

7. Assimilation in high-dimensional model

7.1. MICOM model and assimilation problem

In this section, we show how the AF can be designed in a simple way to produce the high performance estimates for the ocean state in the high-dimensional ocean model MICOM. For details on the Miami Isopycnal Coordinate Ocean Model (MICOM) used here, see Ref. [24]. The model configuration is a domain situated in the North Atlantic from 30°N to 60°N and 80°W to 44°W; for the exact model domain and some main features of the ocean current (mean, variability of the SSH, velocity) produced by the model, see Ref. [24]. The grid spacing is about 0.2° in longitude and in latitude, requiring $N_h = II \times JJ = 25200$ ($II = 140$, $JJ = 180$) horizontal grid points. The number of layers in the model is $KK = 4$. It is configured in a flat bottom rectangular basin (1860 km × 2380 km × 5 km) driven by a periodic wind forcing. The model relies on one prognostic equation for each component of the horizontal velocity field and one equation for mass conservation per layer. We note that the state of the model is $x := (h, u, v)$ where $h = h(i, j, lr)$ is the thickness of lrth layer; $u = u(i, j, lr), v = v(i, j, lr)$ are two velocity components. The layer stratification is made in the isopycnal coordinates, that is, the layer is characterized by a constant potential density of water. The model is integrated from the state of rest during 20 years. Averaging the sequence of states over 2 years 17 and 18 gives a so-called *climatology.* During the period of 2 years 19 and 20, every 10 days (10 ds), we calculate the sea surface height (SSH) from the layer thickness h which will serve as a source for generating observations to be used in the assimilation experiments (in total, there are 72 observations).

7.2. Different filters

The filter used for assimilating SSH observations is of the form

$$\hat{x}_{k+1} = F[\hat{x}_k] + K\zeta_{k+1}, k = 0, 1, \ldots \tag{7.1}$$

where $\hat{x}_{k+1}$ is the filtered estimate for x_{k+1}, $x_{k+1} = [h_{k+1}, u_{k+1}, v_{k+1}]$ is the system state at $(k+1)$ assimilation instant, $F(.)$ represents the integration of the nonlinear MICOM model over 10 days, K is the filter gain, ζ_{k+1} is the innovation vector. The gain K is of the form (**4.11**) where the ECM M will be estimated from the MICOM model. In the experiment, to be closed to realistic situations, only SSH at the grid points $i = 1, \ldots, 140$, $j = 1, .., 180$ are collected as observations. Thus, the observations are available not at all model grid points. The gain K is symbolically written as $K = (K_h, K_u, K_v)^T$ with K_u, K_v representing the operators which produce the correction for the velocity (u, v) from the layer thickness correction $K_h\zeta$ using the geostrophy hypothesis. The filter thus is a reduced order which has the gain K_h to be estimated from S-PE samples.

7.2.1. *PEF: computation of ECM*

In the experiment, two assimilation methods will be implemented. First the PEF is designed. To do that, the data ECM M_d (see Eqs. (**6.5**) and (**7.4**), below) is performed by generating an ensemble of PE samples (as done in the experiment with Lorenz system, see the sampling procedure in Ref. [23] for more detail). As the number of elements of ECM is of order 10^{10} (for only the layer thickness component h), it is impossible to simulate a sufficient number of PE samples so that M_d would be a good estimate for the ECM. The matrix M_d will be used only as data to estimate the parameters in a parametrized ECM as follows (see Ref. [18]):

Let $M \in R^{n_h \times n_h}$ be the ECM for the layer thickness h, that is, $M = M(\mathbf{s}, s')$. One useful and efficient way to simplify the filter structure is to assume that the ECM M has a SeVHS. Assuming there exist two covariance matrices, M_v and M_h such that

$$M(s,s') = M_v(s_v, s_{v'}) \otimes M_h(s_h, s_{h'}), s_v := l, s_h := (i,j), \tag{7.2}$$

where $\otimes$ denotes the Kronecker product between two matrices [25],

$$M_v(s_v, s_{v'}) \otimes M_h(s_h, s_{h'}) = M(i,j,l;i',j',l') = \begin{pmatrix} m_v(1,1)M_h & m_v(1,2)M_h & \dots & m_v(1,n_v)M_h \\ m_v(2,1)M_h & m_v(2,2)M_h & \dots & m_v(2,n_v)M_h \\ \dots & \dots & \dots & \dots \\ m_v(n_v,1)M_h & m_v(n_v,2)M_h & \dots & m_v(n_v,n_v)M_h \end{pmatrix} \tag{7.3}$$

The main advantage of the separability hypothesis is that the number of parameters to be estimated in the covariance matrix is reduced drastically. As a consequence, even an ensemble of PE samples with small size can serve as a large data set for estimating the unknown parameters. This results in a fast convergence of the estimation procedure. In addition, introducing the SeVHS hypothesis allows to avoid the rank deficiency problem in the estimation of the ECM. In fact, as only a few numbers of ScVs can be computed in very high-dimensional systems, approximation of the ECM M by Eq. (**6.5**) results in rank deficiency for M. With such an ECM, the resulting filter will probably produce worse results, not to say on instability which may occur during the filtering process.

Suppose we are given the ensemble of S-PE samples $S_\tau[L] = [\delta h_\tau^{(1)}, \dots, \delta h_\tau^{(L)}]$ which are obtained at the τ time instant by applying the sampling procedure in Ref. [23] subject to L perturbations. For $\tau = 1, \dots, T$, the ECM M_d in Eq. (**4.12**) is estimated as

$$\begin{aligned} M_d &= \frac{1}{T} \sum_{\tau=1}^{T} M_\tau, \\ M_\tau &:= \frac{1}{L} S_\tau(L) S_\tau^T(L) \end{aligned} \tag{7.4}$$

For the problem Eqs. (**4.3**)–(**4.4**) let us define the vector of unknown parameters in the ECM $M(s,s')$ as

$$\theta := (c_{11},\dots,\ c_{1n_v}, c_{21},\dots,\ c_{2n_v}, c_{n_v1},\dots,\ c_{n_vn_v}, L_d)^T, c_{kl} := m_v(k,l). \tag{7.5}$$

where $c_{kl} := m_v(k,l)$. As to the parameter L_d, it represents the correlation length of the horizontal ECM M_h,

$$M_h(y,y') = exp(-d(y,y')/L_d), \tag{7.6}$$

$d(y,y')$ is the distance between two horizontal points $y := (i,j)$ and $y' := (i',j')$.

Considering M_d as data matrix, optimization problem for determining the vector θ looks like

$$\begin{aligned} J[\theta] &= E[\Psi(M_t,\theta)] \rightarrow \min{}_{\theta}, \\ \Psi(M_t,\theta)] &:= ||M_t - M_v(s_v,s_{v'})\otimes M_h(s_h,s_{h'})||_F^2, \end{aligned} \tag{7.7}$$

Mention that the problem Eq. (**7.4**) is somewhat closely related to the Nearest Kronecker Product (NKP) problem [25]. In the experiment, the correlation length L_d is not estimated and is taken identical in two filters PEF and CHF, $L_d = 25$.

7.2.2. PEF: computation of gain

As to the computation of the gain, introduce the notations: at the instant k, let $x(i,j,l)$ be the value of the system state defined at the grid points (i,j,l). Let $\vec{x} = (\vec{x}_1^T, \vec{x}_2^T,\dots,\ \vec{x}_{n_v}^T)^T$ be a vector representation for x where $\vec{x}_l$ is a vector whose components are the values of x at all the horizontal grid points (ordered in some way) at the lth, vertical layer.

Consider the ECM Eq. (**3.10**) and the observation equation (**1.2**). Represent the observation matrix H in a block-matrix form

$$H = [H_1,\dots,\ H_{n_v}] \tag{7.8}$$

which corresponds to the vector representation $\vec{x}$, that is,

$$H\vec{x} = \Sigma_{v=1}^{n_v} H_v \vec{x}_v \tag{7.9}$$

Compute the gain according to Eq. (**2.4**). We have

$$\begin{aligned} MH^T &= M_v \otimes M_h H^T = [\Sigma_1^T,\dots,\ \Sigma_{n_v}^T]^T, \\ \Sigma_l &= M_h \sum_{k=1}^{n_v} c_{lk} H_k^T = M_h G_{v,l}, \\ MH^T &= M_d G_v, M_d = \text{block diag}\ [M_h,\dots,\ M_h], \\ G_v &= [G_{v,1}^T,\dots,\ G_{v,n_v}^T]^T, G_{v,l} := \sum_{k=1}^{n_v} c_{lk} H_k^T, \\ \Sigma &:= HMH^T + R = \sum_{k=1,l=1}^{n_v} c_{lk} H_l M_h H_k^T + R, \end{aligned} \tag{7.10}$$

As proved in Ref. [18], in this case, the gain has the following form

$$K = M_d K_v, K_v = [K_{v,1}^T,\dots, K_{v,n_v}^T]^T = G_v \Sigma^{-1},$$
$$K_{v,l} = G_{v,l} \Sigma^{-1}, l = 1,\dots, n_v, \tag{7.11}$$

where $G_{v,l}, \Sigma$ are defined in Eq. (**7.9**).

7.2.3. Cooper-Haines filter: CHF

To see the performance of the AF, we implement also a so-called CHF. The CHF [7] is obtained from (**7.9**), (**7.10**) under three hypotheses [24]: (**H1**) the analysis error of the system output is canceled in the case of noise-free observations; (**H2**) conservation of the linear potential vorticity (PV); (**H3**) there is no correction for the velocity at the bottom layer. The AF in Ref. [24] is obtained by relaxing one or several hypotheses (**H1**)–(**H3**). From the filtering theory, the difference between the PEF and CHF is lying in the way we estimate the elements of the ECM.

For the choice of the tuning parameters in the PEF, see Ref. [24].

7.3. Numerical results

First, we run the model initialized by the climatology. This run is different from that used for modeling the sequence of true ocean states only by changing the initial state by climatology. This run is denoted as *model*. **Figure 6** shows the (spatial) averaged variance between SSH observations and that produced by the *model*. We see the error grows as time progresses, meaning instability of the numerical model wrt the perturbed initial condition. That fact signifies that the VM will have difficulties in producing high performance estimates if the assimilation window is long.

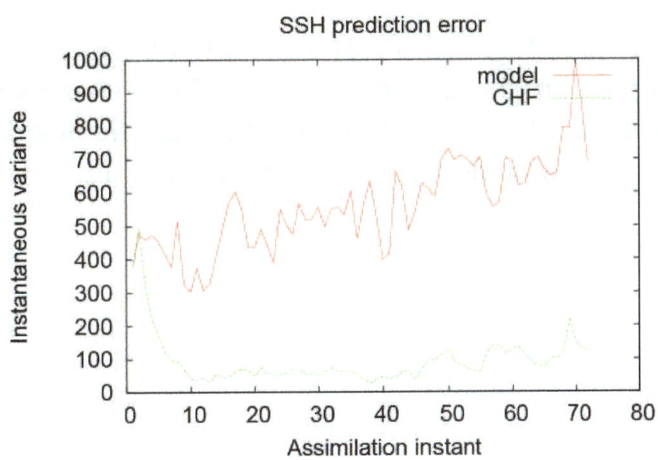

Figure 6. SSH prediction errors resulting from "model" and CHF: growing of the prediction error in the "model" signifies instability of the numerical model wrt specification of the initial system state.

Next the two filters, CHF and PEF, are implemented under the same initial condition as those carried out with the experiment *model*. It is seen from **Figure 6** that initialized by the same initial condition, and the CHF is much more efficient than the *model* in reducing the estimation error. The performance comparison between the CHF and PEF is presented in **Figure 7**. Here, the (spatially averaged variance) SSH prediction errors, resulting from two filters CHF and PEF, are displayed. The superiority of the PEF over the CHF is undoubted. It is clear from **Figure 7** that the PEF is capable of producing the better estimates, with lower error level, along all assimilation period. On the other hand, if the estimation error in CHF decreases continuously at the beginning of the assimilation and is stabilized more or less after (during the interval $k \in (10 : 45)$), it becomes to increase considerably at the end of the assimilation period. It means that the PEF is more efficient than the CHF and the fact that the ECM in the PEF, constructed on the basis of the S-PE samples, has the effect to stabilize its behavior.

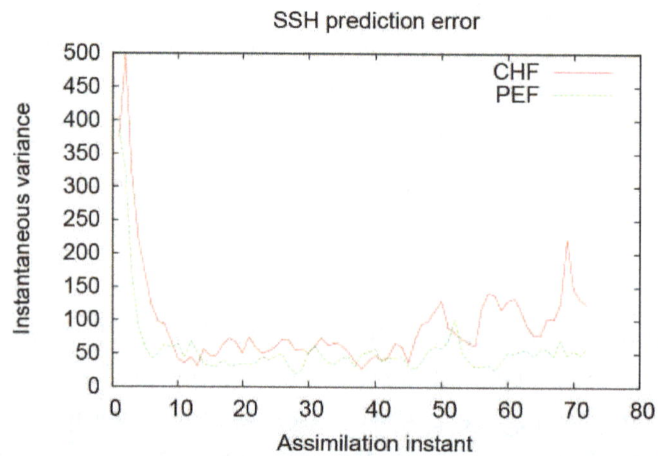

Figure 7. SSH prediction errors resulting from two filters: CHF and PEF. The PEF is capable of producing the estimates with lower estimation errors and is stable along all assimilation period, whereas the CHF has a difficulty to maintain the same performance at the end of the assimilation period. The PEF is much better than the CHF in providing better estimates for the system states.

To see the effect of adaptation, **Figure 8** displays the filtered errors for the *u*-velocity component estimates at the surface, produced by the PEF and APEF, respectively. The APEF is an AF, which is an adaptive version of the PEF. Here, the tuning parameters are optimized by the SPSA method. From the computational point of view, the SPSA requires much less time

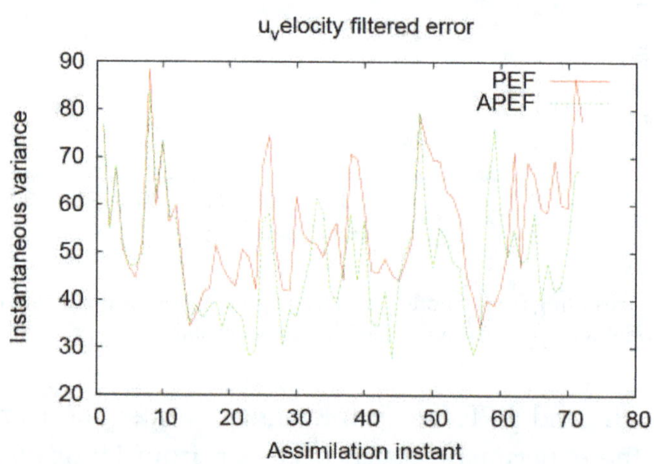

Figure 8. Filtered errors for the *u*-velocity component estimate, resulting from PEF and APEF (AF based on PEF). Optimization is performed by SPSA. By tuning the parameters in the filter gain, one can improve considerably the performance of the PEF.

integration and memory storage compared with the traditional AE method. At each assimilation instant, we have to make only two integrations of the MICOM for approximating the gradient vector. From **Figure 8**, one sees that the adaptation allows to reduce significantly the estimation errors produced by the PEF.

8. Conclusions

In this chapter, a comparison study on the performance of the AF with other existing methods is presented in the context of its application to data assimilation problems in high-dimensional numerical models. As it is seen, in comparison with the standard VM, the AF is much simpler to implement and produces better estimates. The advantages of the AF over other methods such as EKF, CHF are also demonstrated. The principal reason for high performance of the AF is lying in the choice of innovation representation for the initial input-output system and selection of pertinent gain parameters as control variables to minimize the MSE of the innovation process. If in the VM, the choice of the structure for the initial state is the most important thing to do (but that is insufficient for guaranteeing its high performance), in the AF, however, the initial state has a little impact on the performance of the AF. This happens because the AF is selected as optimal in the class of stable filters, and as a consequence, the error in the initial estimate is attenuated during assimilation process. In contrary, in the VM, the error in the specification of the initial guess is amplified during assimilation if the numerical model is unstable.

In conclusion, it is important to emphasize that the AF approach, presented in this chapter, is consolidated by exploiting the following road map: (i) generate data ECM from S-PE samples which grow in the directions of dominant Schur vectors; (ii) select a parametrized structure for the ECM under the hypothesis SeVHS; (iii) make the choice of tuning parameters in the gain by minimizing the distance between the data ECM and that having the SeHVS structure; (iv) adjust the unknown parameters in the gain in order to minimize the PE error of the system output by applying the SPSA algorithm.

There are a wide variety of engineering problems to which the AF is applicable and that could be worthy of further study. Depending on particular problems, undoubtedly, the other modifications would be helpful to improve the filter performance, to simplify its implementation. But the main features of the AF, presented in this chapter, remain as the key points to follow in order to preserve a high performance of the AF.

Author details

Hong Son Hoang* and Remy Baraille

*Address all correspondence to: hhoang@shom.fr

SHOM/HOM/REC, Toulouse, France

References

[1] Bucy R.S. and Joseph D.P.: Filtering for Stochastic Processes, with Applications to Guidance. New York: Wiley; 1968.

[2] Maybeck P.S.: Stochastic Models, Estimation, and Control. Vol. 1, Academic Press; 1979.

[3] Kalman R.E.: A new approach to linear filtering and prediction problems. Trans ASME J. Basic Eng. 1960; 82: 35–45.

[4] Todling R. and Cohn S.E.: Suboptimal schemes for atmospheric data assimilation based on the Kalman filter. Monthly Weather Rev. 1994; **122**: 2530–2557.

[5] Ghil M. and Manalotte-Rizzoli P.: Data assimilation in meteorology and oceanography. Adv Geophys. 1991; **33**: 141–266.

[6] Simon D.: Optimal State Estimation. Hoboken, NJ: John Wiley and Sons; 2006. ISBN 978-0-471-70858-2.

[7] Cooper M. and Haines K.: Altimetric assimilation with water property conservation. J Geophys Res. 1996; **101**: 1059–1077.

[8] Nocedal J. and Wright S.: Numerical Optimization. New York: Springer-Verlag; 1999. ISBN 0-387-9873-2.

[9] Talagrand O.: Data assimilation problems. In: Raschke E. and Jacob D., editors. Energy and Water Cycles in the Climate System. NATO ASI Series I, Vol. **5**; 1993, pp. 187–213.

[10] Hoang H.S., De Mey P., Talagrand O. and Baraille R.: A new reduced-order adaptive filter for state estimation in high dimensional systems. Automatica. 1997; **33**: 1475–1498.

[11] Gevers M. and Kailath T.: An innovations approach to least-squares estimation, Pt. VI: discrete-time innovations representations and recursive estimation. IEEE Trans Autom Control. 1973; 18(6), **12**: 588–600.

[12] Hoang H.S., Talagrand O. and Baraille R.: On the design of a stable adaptive filter for high dimensional systems. Automatica. 2001; **37**, 341–359.

[13] Baram Y. and Kalit G.: Order reduction in linear state estimation under performance constraints. IEEE Trans Autom Control. 1987; AC-32, **11**: 983–989.

[14] Spall J. C. Accelerated second-order stochastic optimization using only function measurements. In: Proceedings of 36th IEEE Conference on Decision and Control, 14171424; 1997.

[15] Hoang H.S. and Baraille R.: Stochastic simultaneous perturbation as powerful method for state and parameter estimation in high dimensional systems. In: Albert R. Baswell, editors. Advances in Mathematics Research. Vol 20, Nova Science Publishers.; 2015, pp. 117–148.

[16] Evensen G.: The ensemble Kalman filter: theoretical formulation and practical implementation. Ocean Dyn. 2003; **53**: 343–367.

[17] Hamill T.M.: Ensemble-based atmospheric data assimilation. In: Palmer T., editor. Predictability of Weather and Climate. Cambridge Press; 2006; pp. 124–156.

[18] Hoang H.S. and Baraille R.: A low cost filter design for state and parameter estimation in very high dimensional systems. In: Proceedings of the 19th IFAC Congress; August, 2014, pp. 3256–3261.

[19] Daley R.: The effect of serially correlated observation and model error on atmospheric data assimilation. Monthly Weather Rev. 1992; **120**: 165–177

[20] Baraille R.: Modélisation numérique avec le modèle HYCOM au SHOM. At "http://mathocean.math.cnrs.fr/presentations/Baraille.pdf"

[21] Lorenz E.N.: Deterministic non-periodic flow. J Atmos Sci. 1963; **20**: 130–141.

[22] Kivman G.A.: Sequential parameter estimation for stochastic systems. Nonlinear Process Geophys. 2003; **10**: 253–259. doi:10.5194/npg-10-253-2003

[23] Hoang H.S. and Baraille R.: Prediction error sampling procedure based on dominant Schur decomposition. Application to state estimation in high dimensional oceanic model. J Appl Math Comput. 2011; **12**: 3689–3709.

[24] Hoang H.S., Baraille R. and Talagrand O.: On an adaptive filter for altimetric data assimilation and its application to a primitive equation model MICOM. Tellus, 57A, 2005, **2**: 153–170.

[25] Golub G.H. and Van Loan C.F.: Matrix Computations. Johns Hopkins University Press; 1996.

9

Nonlinear Transformations and Radar Detector Design

Graham V. Weinberg

Abstract

A nonlinear transformation is introduced, which can be used to compress a series of random variables. For a certain class of random variables, the compression results in the removal of unknown distributional parameters from the resultant series. Hence, the application of this transformation is investigated from a radar target detection perspective. It will be shown that it is possible to achieve the constant false alarm rate property through a simple manipulation of this transformation. Due to the effect the transformation has on the cell under test, it is necessary to couple the approach with binary integration to achieve reasonable results. This is demonstrated in an X-band maritime surveillance radar detection context.

Keywords: transformations, random variable properties, radar detection, mathematical statistics, radar

1. Introduction

The fundamental problem to be examined in this chapter is the detection of targets embedded within the sea surface, from an airborne maritime surveillance radar. Artifacts of interest could be lifeboats or aircraft wreckage resulting from aviation or maritime disasters. From a military perspective, one may be interested in the detection and tracking of submarine periscopes. Another scenario may be the detection of illegal fishing vessels or small boats used for smuggling of people or contraband. An airborne maritime surveillance radar has a difficult task in the detection of such objects from high altitude, while surveying a very large surveillance volume.

Such radars operate at X-band and are high resolution, and as such are affected by backscattering from the sea surface, which is referred to as clutter. This backscattering tends to mask small targets and makes the surveillance task extremely difficult. One of the major issues with the design of radar detection schemes is the minimization of the detection of false targets, while maximizing the detection of real targets. As a statistical hypothesis test, one can apply the Neyman-Pearson Lemma to produce a decision rule that achieves these objectives. However, in many cases, such a decision rule requires clutter model parameter approximations as well as estimates of the target strength based upon sampled returns. An issue, well known within the radar community, is that small variations in the clutter power level can result in huge increases in the number of false alarms. Since clutter power is a function of the underlying clutter model's parameters, approximations of the latter will have an inevitable effect on the former. Hence a large body of research has been devoted to designing radar detection strategies that maintain a fixed level of false alarms. A detector that achieves this objective is said to have the constant false alarm rate (CFAR) property [1].

In order to maintain a fixed rate of false alarms, sliding window decision rules were examined in early studies of radar detection strategies [2–6]. These investigations have been extended to account for different clutter models and to address issues with earlier detector design in a number of subsequent analyses [7–15]. Such decision rules can be formulated as follows. Suppose that the statistic Z is the return to be tested for the presence of a target. Let $Z_1, Z_2, \ldots, Z_N$ be N statistics from which a measurement of the level of clutter is taken, via some function $f = f(Z_1, Z_2, \ldots, Z_N)$. Then a target is declared present in the case where Z is larger than a constant times f. The constant is selected so that in ideal scenarios, the false alarm rate remains fixed. It is generally assumed that the clutter statistics are independent and identically distributed in ideal settings, and also independent of the statistic Z. This can be formulated as a statistical hypothesis test by letting H_0 be the hypothesis that the cell under test (CUT) statistic Z does not contain a target, and H_1 the alternative that it contains a target embedded within clutter. Then the test is written

$$Z \underset{H_0}{\overset{H_1}{\gtrless}} \tau f(Z_1, Z_2, \ldots, Z_N) \tag{1}$$

where $\tau > 0$ is the threshold constant and the notation used in Eq. (1) means that H_0 is rejected when $Z > \tau f(Z_1, Z_2, \ldots, Z_N)$. The probability of false alarm is given by

$$\text{Pfa} = \mathbb{P}\left(Z > \tau f(Z_1, Z_2, \ldots, Z_N) \middle| H_0\right). \tag{2}$$

If τ can be determined, for a specified Pfa in Eq. (2), such that it is independent of clutter parameters, then the decision rule in Eq. (1) will be able to maintain the CFAR property in ideal scenarios. In practical radar systems, a detection scheme such as in Eq. (1) can be run across the data returns sequentially to allow binary decisions on the presence of targets to be made, which are then passed to a tracking algorithm. A comprehensive examination of such detection processes is included in [1].

This chapter examines an alternative approach to achieve the CFAR property, based upon a nonlinear transformation that is used to compress the original clutter sequence. The consequence of this is that the resulting transformed series of random variables will have a fixed clutter power level and so permits a CFAR detector to be proposed. It is then shown how this transformation can be used to produce a practical radar detection scheme.

The chapter is organized as follows. Section 2 introduces the nonlinear mapping and formulates a decision rule. Section 3 specializes this to the case of Pareto distributed sequences, since the Pareto model is suitable for X-band maritime surveillance radar clutter returns. Section 4 demonstrates detector performance in homogeneous clutter, while Section 5 applies the decision rules directly to synthetic target detection in real X-band radar clutter.

2. Transformations and decision rule

2.1. Mapping

In X-band maritime surveillance radar, the Pareto distribution has become of much interest as a clutter intensity model due to its validation relative to real radar clutter returns [16–18]. This model arises as the intensity distribution of a compound Gaussian model with inverse Gamma texture. Consequently, the Pareto distribution fits into the currently accepted radar clutter model phenomenology [19]. Hence, there have been a number of recent advances in the design of CFAR processes under a Pareto clutter model assumption [20–25].

A random variable X has a Pareto distribution [26] with shape parameter $\alpha > 0$ and scale parameter $\beta > 0$ if its cumulative distribution function (cdf) is

$$F_X(t) := \mathbb{P}(X \leq t) = 1 - \left(\frac{\beta}{t}\right)^{\alpha}, \tag{3}$$

for $t \geq \beta$. The density of X follows by differentiation of Eq. (3). In order to ensure the existence of the first two moments, it is usually assumed that $\alpha > 2$, which is an assumption that has been validated in fits of this model to real data [18]. This Pareto model possesses what is referred to as a duality property in Ref. [20]. To introduce this, recall that if Y is an Exponential random variable with unity mean, its cdf is given by

$$F_Y(t) = 1 - e^{-t} \tag{4}$$

for $t \geq 0$. Then it can be shown that the Pareto model in Eq. (3) can be related to Eq. (4) via the random variable relationship

$$X = \beta e^{\alpha^{-1} Y}. \tag{5}$$

Other random variables of interest in radar signal processing, such as the Weibull, can also be expressed in a form similar to Eq. (5). Hence, for the purposes of generality, suppose $\{X_j, j \in \mathbb{N} := \{0, 1, 2, \ldots\}\}$ is a sequence of homogeneous random variables with common support and that θ_1 and θ_2 are two fixed real constants. Define a sequence of random variables $\{Z_j, j \in \mathbb{N}\}$ by

$$Z_j = \theta_1 X_j^{\theta_2}. \tag{6}$$

The sequence produced via Eq. (6) is a generalization of the Pareto model (3). Next define a nonlinear mapping $\zeta : \mathbb{R}^+ \times \mathbb{R}^+ \times \mathbb{R}^+ \times \mathbb{R}^+ \rightarrow \mathbb{R}^+ \cup \{0\}$ by

$$\zeta(x_1, x_2, x_3, x_4) = \left|\frac{\log(x_1) - \log(x_2)}{\log(x_3) - \log(x_4)}\right|, \tag{7}$$

where each $x_j > 0$, $x_3 = x_4$ and $\mathbb{R}^+$ is the positive real numbers. Then the following result is relatively easy to prove:

Lemma 2.1 *Suppose* $\{Z_j,\ j \in \mathbb{N}\}$ *is a sequence of random variables defined via Eq. (6). Then the sequence* $\{W_j, j \in \mathbb{N}\}$ *with* $W_j := \zeta(Z_j, Z_{j+1}, Z_{j+2}, Z_{j+3})$ *does not depend on* θ_1 *and* θ_2.

The proof of Lemma 2.1 is now outlined. Supposing that Z_j, Z_{j+1}, Z_{j+2} and Z_{j+3} are represented in the form defined via Eq. (6) it follows that

$$\begin{aligned} W_j &= \left| \frac{\log(\theta_1 X_j^{\theta_2}) - \log(\theta_1 X_{j+1}^{\theta_2})}{\log(\theta_1 X_{j+2}^{\theta_2}) - \log(\theta_1 X_{j+3}^{\theta_2})} \right| \\ &= \left| \frac{\theta_2 \log(X_j) - \theta_2 \log(X_{j+1})}{\theta_2 \log(X_{j+2}) - \theta_2 \log(X_{j+3})} \right| \\ &= \left| \frac{\log(X_j) - \log(X_{j+1})}{\log(X_{j+2}) - \log(X_{j+3})} \right|, \end{aligned} \tag{8}$$

where properties of the logarithmic function have been utilized. Since Eq. (8) does not depend on θ_1 and θ_2, the proof is completed.

Lemma 2.1 suggests that if the original sequence of random variables is processed in 4-tuples, the compressed sequences' statistical structure is only dependent on the random variables X_j. Observe that the Lemma does not require an independence assumption. Thus if sequence $\{X_j\}$ has no unknown statistical parameters, the process generated by Eq. (7) also has no unknown parameters. This suggests that processing of a data sequence in terms of 4-tuples may be an effective may in which to achieve the CFAR property. The next subsection clarifies this.

2.2. Decision rule

In order to propose a decision rule exploiting the transformation introduced in Lemma 2.1, it is necessary to focus first on a series of four returns. Hence, suppose we have a CUT statistic Z, and three clutter measurements are available, denoted Z_1, Z_2 and Z_3. Let H_0 be the hypothesis that the CUT contains no target, and H_1 the hypothesis that it does contain a target embedded within clutter. Then, based upon Eq. (7), a linear threshold test takes the form

$$\zeta(Z, Z_1, Z_2, Z_3) \underset{H_0}{\overset{H_1}{\gtrless}} \tau, \tag{9}$$

where $\tau > 0$ is the threshold. Based upon Lemma 2.1 if the clutter is modelled by Eq. (6), then it is clear that under H_0, the Pfa of the test in Eq. (9) will not depend on θ_1 or θ_2, implying it is CFAR with respect to these parameters. Furthermore, an auxiliary motivation for defining a linear threshold detector such as Eq. (9) is that in the cases where it is assumed that one has *a priori* knowledge of clutter parameters, linear threshold detectors are ideal, or asymptotically optimal, and hence provide the maximum probability of detection within the class of sliding window decision rules [27].

The test in Eq. (9) can also be re-expressed in terms of the preprocessed clutter statistics. In particular, it can be shown to be equivalent to rejecting H_0 if

$$\begin{aligned} Z &> Z_1 e^{\tau|\log(Z_2)-\log(Z_3)|} \\ &\text{or} \\ Z &> Z_1 e^{\tau|\log(Z_2)-\log(Z_3)|} \end{aligned} \tag{10}$$

with the appropriate choice for τ, which can be determined from the corresponding Pfa expression for Eqs. (9) or (10).

Observe that this test is not of the usual form found in the radar signal processing literature, since it compares a CUT with a measurement of clutter based upon three statistics, and not upon a sample of predetermined size. This will be discussed subsequently in terms of practical implementation of the test in Eq. (10). The next section discusses the application of Eq. (10) to the Pareto clutter case, enabling the determination of τ.

3. Specialization to the Pareto Clutter model

3.1. Distributions under H_0

Since the motivation of the work developed here is the design of radar detection schemes for maritime surveillance radar, the results of the previous section are specialized to the Pareto case. In order to apply Lemma 2.1, it is necessary to determine the distribution of the resultant sequence produced by ζ under H_0. The following is the key result:

Corollary 3.1 *In the case where the sequence of random variables in Lemma 2.1 is Pareto distributed and independent, the cdf of the sequence processed by ζ is given by*

$$F_P(t) = \frac{t}{t+1}, \tag{11}$$

for $t \geq 0$.

This can be recognized as a Pareto distribution, with support the nonnegative real line and shape and scale parameter unity. More specifically, $P = X + 1$, where X has density (3) with $\alpha = \beta = 1$. This illustrates the cost of the nonlinear transformation approach: although the resultant series of clutter has no unknown clutter parameters, it is from a distribution with no finite moments. The independence assumption is adopted for analytical tractability and is consistent with the assumption that independent and identically distributed clutter returns are available, as in the formulation of the test in Eq. (1).

To prove Corollary 3.1, suppose that η_1 and η_2 are two independent random variables with cdf (4). Then by analyzing the difference $\eta_1 - \eta_2$, it can be shown that it has cdf

$$F_{\eta_1-\eta_2}(t) = \begin{cases} 1-\frac{1}{2}e^{-t}, & \text{for } t \geq 0 \\ \frac{1}{2}e^{t} & \text{for } t < 0, \end{cases} \tag{12}$$

which is that of a Laplace distribution. Then it follows that

$$F_{|\eta_1-\eta_2|}(t) = \mathbb{P}(-t \le \eta_1 - \eta_2 \le t) = 1 - e^{-t}, \tag{13}$$

where Eq. (12) has been applied, and $t > 0$. Thus the modulus of the difference is also exponentially distributed with unit mean.

Supposing that κ_1 and κ_2 are two independent random variables with cdf Eq. (13), then by statistical conditioning

$$F_{(\kappa_1/\kappa_2)}(t) = \mathbb{P}(\kappa_1 \le t\,\kappa_2) = \int_0^\infty \mathbb{P}(\kappa_1 \le t\omega)\, e^{-\omega}\, d\omega, \tag{14}$$

and an application of Eqs. (4)–(14) shows that the ratio has cdf Eq. (11) with an evaluation of the integral. This establishes the result in Corollary 3.1, as required.

3.2. Thresholds and the CUT

Based upon Corollary 3.1 the univariate threshold for the Pareto case is given by

$$\tau = \mathrm{Pfa}^{-1} - 1. \tag{15}$$

The threshold (Eq. (15)) illustrates the issues with the nonlinear mapping, as this threshold will be quite large for appropriate Pfa. Note that for a Pfa of 10^{-6}, $\tau = 10^6 - 1$. This threshold will increase as the Pfa decreases. In the Pareto setting, it is shown in Ref. [20] that an ideal detector has its threshold set via $\beta(\mathrm{Pfa})^{-1/\alpha}$. In the case where $\alpha = 4.7241$ and $\beta = 0.0446$ (which correspond to spiky clutter returns) and with the Pfa set to 10^{-6}, this threshold is 0.8312 by contrast. Thus the nonlinear mapping, in the process of compressing the original data series, can be used to achieve the CFAR property with Eq. (10), but detection performance may be unacceptable.

To explore this further, it is informative to examine the detection scheme in Eq. (10) when there is a target model present. Suppose Ξ is the CUT statistic, in the case where a target is present in the clutter, in the pretransformed data. Let $\hat{\Xi}$ be the CUT in the transformed domain, meaning the detector Eq. (10) when there is a target present so that T is the intensity measurement of a return signal and clutter in the complex domain. Then by applying the left-hand expression for Pareto random variables in Eq. (5), we can write

$$\hat{\Xi} = \alpha \left| \frac{\log(\Xi/\beta) - E_1}{E_2 - E_3} \right|, \tag{16}$$

where each E_j is an independent exponentially distributed random variable with unit mean. Then with an application of results from the proof of Corollary 3.1, since $|E_2 - E_3|$ has the same exponential distribution, one can apply statistical conditioning on E_1 and $|E_2 - E_3|$ to show that the distribution function of the transformed CUT is

$$\begin{aligned} F_{\hat{\Xi}}(t) &= \int_0^\infty \int_0^\infty e^{-\theta} e^{-\varphi}\, \mathbb{P}\left(|\log(\Xi/\beta) - \theta| \le \frac{\varphi t}{\alpha} \right) d\theta d\phi \\ &= \int_0^\infty \int_0^\infty e^{-\theta} e^{-\varphi}\, \mathbb{P}\left(\beta e^{\theta - \frac{\varphi t}{\alpha}} \le \Xi \le \beta e^{\theta + \frac{\varphi t}{\alpha}} \right) d\theta d\phi \\ &= \int_0^1 \int_0^1 \mathbb{P}\left(\beta x^{-1} y^{\frac{t}{\alpha}} \le \Xi \le \beta x^{-1} y^{-\frac{t}{\alpha}} \right) dx dy \\ &= \int_0^1 \int_0^1 F_{\Xi}\left(\beta x^{-1} y^{-\frac{t}{\alpha}}\right) - F_{\Xi}\left(\beta x^{-1} y^{\frac{t}{\alpha}}\right) dx dy, \end{aligned} \tag{17}$$

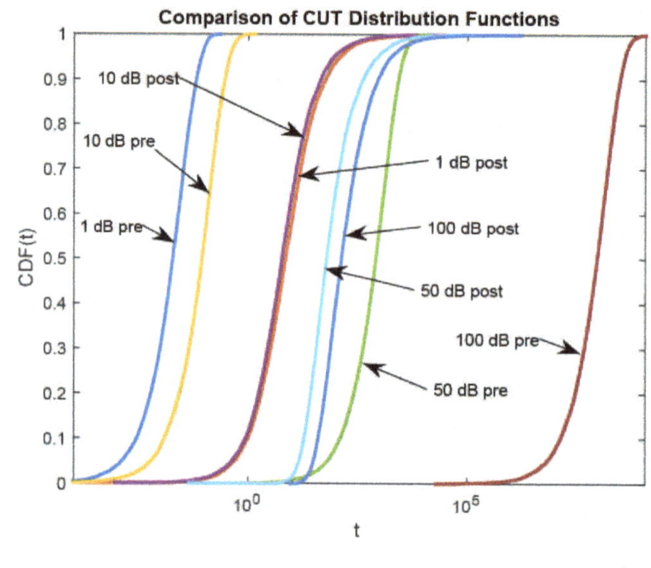

Figure 1. Comparison of CUT for the pretransformed data (denoted pre) and data processed via the nonlinear mapping (denoted post). The CUT is plotted for a Swerling 1 target model with a given SCR as indicated.

where the change of variables $x = e^{-\theta}$ and $y = e^{-\phi}$ has been applied. Thus the transformed CUT can be generated from the pretransformed CUT via Eq. (17). To examine this, **Figure 1** plots Eq. (17) in the case of a Swerling 1 target model embedded within Pareto distributed clutter with $\alpha = 4.7241$ and $\beta = 0.0446$. A Swerling I target model is essentially a bivariate Gaussian model, which is combined with the Pareto model by embedding the latter into a compound Gaussian process with inverse Gamma texture in the complex domain, and then taking modulus squared to produce the intensity measurement [20]. The distribution function of Ξ can also be found in Ref. [20] for the case of interest. **Figure 1** shows the pretransformed CUT as well as Eq. (17), in the cases where the signal to clutter (SCR) ratio is 1, 10, 50, and 100 dB. For the case of a 1 dB target model, the CUT has its range of potential values increased under the transformation. This is also the same for the 10 dB case. Interestingly, for the 50 dB and 100 dB cases, the situation is reversed. Hence, as the SCR increases, the nonlinear mapping suppresses the target SCR, reducing the range of admissible values for the transformed CUT. This suggests that although the nonlinear mapping removes unknown clutter parameters, it may also impede detection due to target suppression. If the threshold is set via Eq. (15), then it is clear from **Figure 1** that it will be very difficult to detect targets with a reasonably small Pfa. Hence, the new detection scheme must be combined with an integration process to rectify this.

4. Performance in homogeneous clutter

4.1. Methodology and data

In order to examine the performance of the proposed detection scheme (9), clutter is simulated under the assumption of a Pareto clutter model, which has been found to fit Defence Science and Technology Group's (DSTG's) real X-band maritime surveillance radar data sets. Ingara is an experimental X-band imaging radar which has provided real clutter for the analysis of detector performance [28]. A trial in 2004 produced a series of clutter sets that have been analyzed from a statistical perspective in Ref. [29]. During the trial, the radar operated in a circular spotlight mode, surveying the same patch of the Southern Ocean at different azimuth and grazing angles. Additionally, the radar provided full polarimetric data. For the purposes of the numerical work to follow, focus is restricted to one particular data set. This is run 34683, at an azimuth angle of 225°, which is approximately in the up wind direction. Additionally, the numerical analysis focuses on the horizontal transmit and receive (HH) case.

For performance analysis in homogeneous clutter, the data is simulated with distributional parameters matched to those obtained from the Ingara data set. The data consists of 821 pulses with 1024 range compressed samples, from which maximum likelihood estimates of the distributional parameters can be obtained from the intensity measurements. Under the Pareto model assumption, the estimates are $\hat{\alpha} = 4.7241$ and $\hat{\beta} = 0.0446$.

As remarked previously, it is necessary to couple (10) with an integration scheme to enhance its performance. The integration scheme used for this purpose is binary integration, which is well-described in Ref. [30], and an application of it in a Pareto distributed clutter environment can be found in Ref. [31]. Such a process applies a series of $M \geq 1$ tests of Eq. (10), and then conclude that if at least S out of M return a detection result, then a target is likely to be present in the radar clutter [30], where $S \in \{1, 2, \ldots, M\}$. Selection of an appropriate S is outlined in Ref. [31]. Essentially, it is pointed out in Ref. [32] that for a specified univariate cumulative detection probability and false alarm rate and a fixed number of maximum binary integration returns M, there exists an optimal S which minimizes the required signal to clutter ratio, and maximizes the binary integration gain. This can be done visually or numerically by plotting the minimum SCR as a function of S, under the assumption of a certain signal model. This approach, and the analysis in Ref. [31], shows that in the current context, the choice of $S = 3$ with $M = 8$ should provide good results. Relative to the problem addressed in this chapter, applying binary integration with a linear threshold detector in the transformed clutter domain is not computationally expensive, and thus is seen as a reasonable solution.

If $\mathrm{Pfa}_{\mathrm{BI}}$ denotes the Pfa for binary integration, then it can be expressed in terms of the univariate detection processes Pfa through the equation

$$\mathrm{Pfa}_{\mathrm{BI}} = \sum_{j=S}^{M} \binom{M}{j} \mathrm{Pfa}^{j} (1 - \mathrm{Pfa})^{M-j}. \tag{18}$$

The threshold τ is set via Eq. (18) coupled with the univariate Pfa from Eq. (9).

To simulate detection performance, the probability of detection (Pd) is estimated, using 10^6 Monte Carlo runs based upon a Swerling 1 target model assumed for the CUT. For each SCR, the binary integration process is run using $S = 3$ out of $M = 8$ binary integration. The motivation for these choices can be found in Ref. [31]. In order to assess the robustness of the detection scheme to interference, up to two interfering targets are inserted into the clutter measurements to give an indication of the performance with interference. Thus independent Swerling 1 targets, with interference to clutter ratio (ICR) of 1 dB, are applied to Z_1 (denoted Inter 1 in the plots), then to Z_2 (denoted Inter 2), and then to both Z_1 and Z_2 (denoted Inter 3) in the univariate decision rule in Eq. (9). A real spurious target may only appear in a subset of the clutter measurements and so this analysis of interference can be viewed as an upper bound on poor performance.

4.2. Receiver operating characteristic curves

Receiver operating characteristic (ROC) curves are used to examine the performance, which plots the probability of detection as a function of the false alarm probability, when the target in the CUT is at a fixed SCR. **Figures 2–4** provide examples of the performance of the new detector Eq. (10) with binary integration and compares it to the performance of some of the recently introduced detectors designed for operation in a Pareto clutter model environment. For a CUT Z and clutter range profile $Z_1, Z_2, \ldots, Z_N$, the Geometric Mean (GM) CFAR is

$$Z \underset{H_0}{\overset{H_1}{\gtrless}} \beta^{1-N\zeta} \prod_{j=1}^{N} Z_j^{\zeta}, \tag{19}$$

which is shown in Ref. [20] to have its threshold set via $\zeta = \text{Pfa}^{-1/N} - 1$. Similarly, an Order Statistic (OS)-CFAR has been analyzed in Ref [22], which is given by

$$Z \underset{H_0}{\overset{H_1}{\gtrless}} \beta^{1-\nu_j} Z_{(j)}^{\nu_j}, \tag{20}$$

which has its threshold multiplier ν_j set via inversion of the Pfa equation given by

$$\text{Pfa} = \frac{N!}{(N-j)!} \frac{\Gamma(\nu_j + N - j + 1)}{\Gamma(\nu_j + N + 1)}, \tag{21}$$

where the OS index $1 \leq j \leq N$ and the notation ν_j emphasizes the fact that ν_j depends on the selected OS index j. Observe that both these decision rules require *a priori* knowledge of β. In order to provide a valid comparison with Eq. (10), these detectors have been applied with $N = 3$ and coupled with binary integration. Due to this, there are three choices available for j, corresponding to a minimum (denoted MIN, when $j = 1$), median (MED, $j = 2$), and maximum (MAX, $j = 3$).

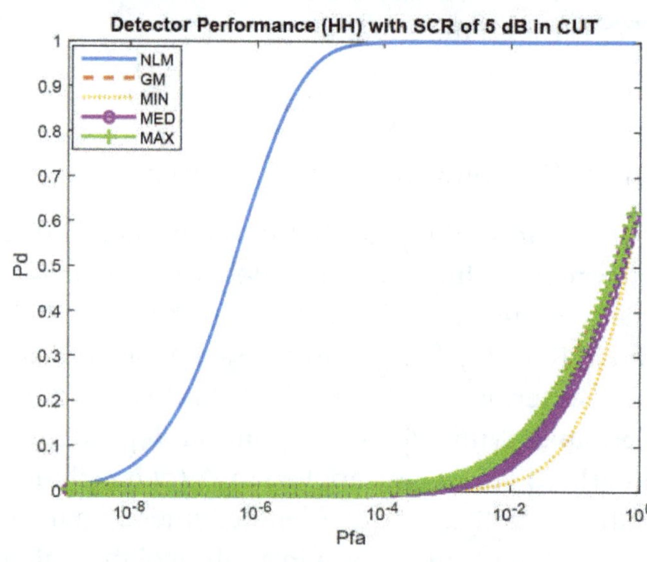

Figure 2. Comparison of detectors with a small target SCR.

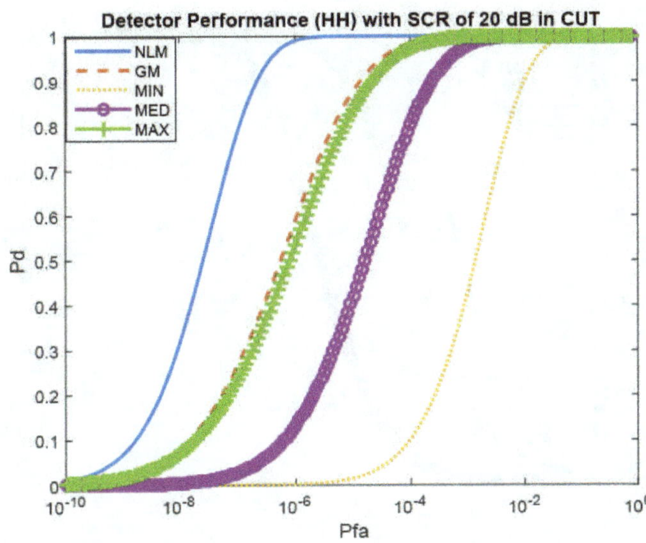

Figure 3. Detector performance with a larger SCR in the CUT.

Figure 2 compares the performance of these decision rules, where the detection process (10) coupled with binary integration is denoted as the nonlinear mapping (NLM). In this case, the CUT SCR is 5 dB, representing a small target. As can be observed, the new decision rule has superior performance. The same experiment is repeated in **Figure 3**, where the CUT SCR is 15 dB, and then it is increased to 20 dB in **Figure 4**. These results show that the new detection process has superior performance, while not requiring *a priori* knowledge of the Pareto scale parameter. These results validate the application of Eq. (10) to target detection in spiky X-band clutter with binary integration.

It is interesting to note that as M is increased, there is very little gain in performance. To demonstrate this, **Figure 5** repeats the same scenario in **Figure 4** except M has been increased to

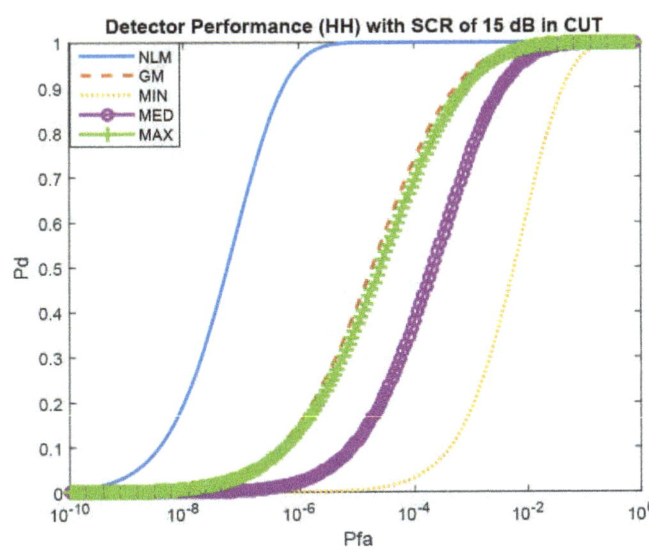

Figure 4. Decision rule performance with a CUT SCR of 20 dB.

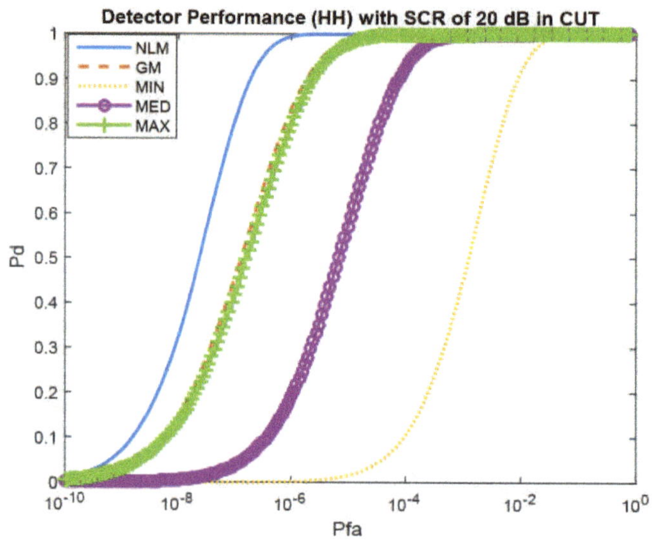

Figure 5. Decision rule performance with a CUT SCR of 20 dB, where the binary integration is $S = 3$ out of $M = 30$.

30. Comparing **Figures 4** and **5**, it is clear that there is very little gain. However, the computational complexity increases dramatically as M is increased. Hence, in a practical implementation of the binary integration process, it is more efficient to select M small.

4.3. Effect of interference

Next the cost of interference on the new decision rule is examined, and for brevity, only this decision rule is considered. **Figure 6** shows the case where the CUT has SCR of 5 dB, and the decision rule (10) coupled with binary integration is denoted BI, while the three interference cases are marked appropriately. Here we observe quite good performance that decreases with the interference. **Figure 7** shows the result of increasing the SCR in the CUT to 20 dB. The result is an expected detection performance improvement as shown.

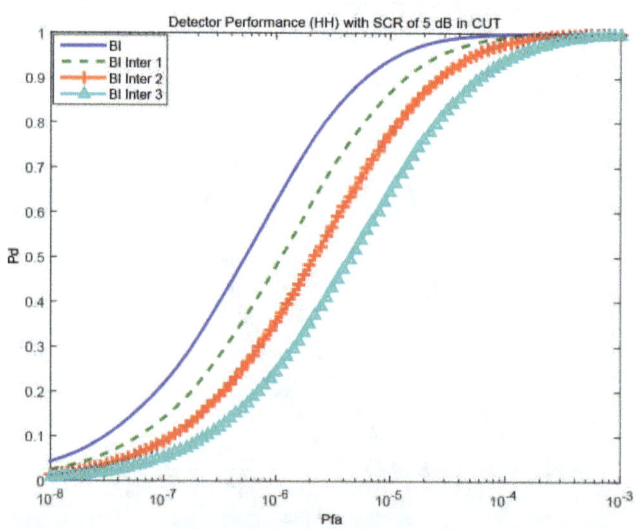

Figure 6. Performance of new detector when subjected to interference.

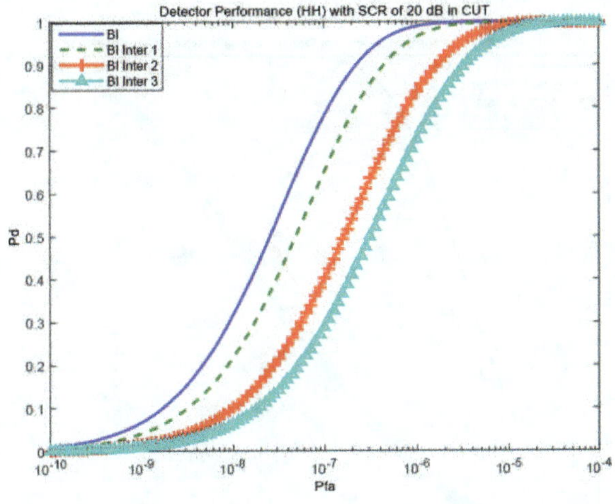

Figure 7. ROC for higher SCR with interference.

5. Performance in real data

As a final test of the proposed detection scheme, it was run directly on the Ingara data set under consideration, with the insertion of synthetic Swerling 1 target and interference as for the homogeneous case. A sliding window was run across the data sequentially, and detection performance was estimated by running the 3 out of 8 detection scheme, resulting in a run length of 840,672. The Ingara data is slightly correlated from cell to cell and so the detector Eq. (9), which has threshold set via an independence assumption, becomes a suboptimal decision rule. Detection performance under both clutter model assumptions is plotted on the same ROC curve to compare performance on the real data more easily. The same scenario is repeated as for the analysis under homogeneous independent clutter.

Figure 8 shows detection performance with the CUT SCR of 5 dB, while **Figure 9** repeats the same numerical experiment as for **Figure 8**, except the CUT has SCR of 20 dB. Comparing

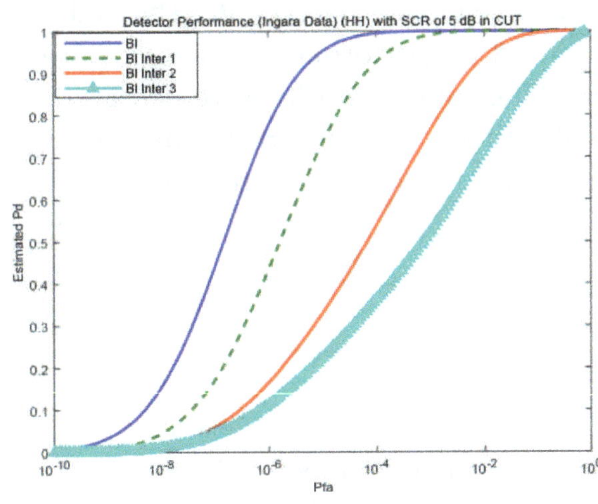

Figure 8. Performance of the detectors on the Ingara data directly.

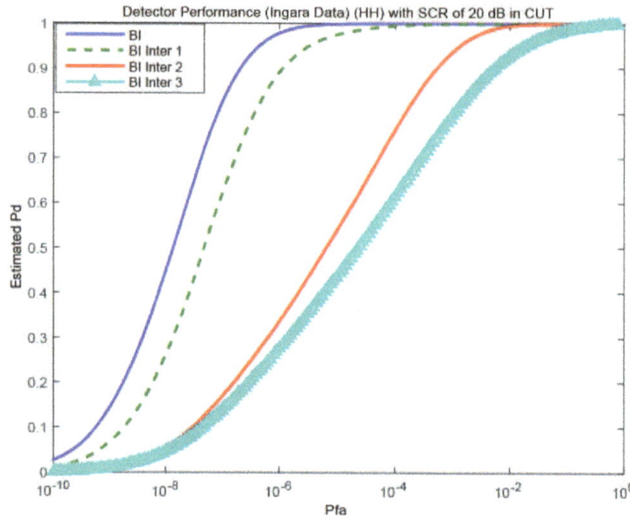

Figure 9. Second example of performance of the detectors on the Ingara data directly.

Figure 8 with **Figure 6** we observe that the effects of correlation are having an effect on the performance in real data. The new decision rule is designed to operate in independent homogeneous clutter returns, and so there is a serious variation in performance. The same situation is observed at a larger CUT SCR (comparing **Figures 9** and **7**).

6. Conclusions

A nonlinear transformation was introduced and shown to remove clutter parameter dependence for a class of statistical models. This was used to formulate a simple linear threshold detector in the transformed clutter domain. Due to issues with the magnitude of detection thresholds, it was necessary to couple the approach with binary integration.

Analysis of detection performance in simulated clutter showed good detection performance. Interference had a strong impact on performance as expected. When the detection process was applied directly to real data, similar results were observed. Nonetheless, the nonlinear transformation, coupled with binary integration, resulted in reasonable detection performance while guaranteeing the CFAR property is preserved.

Author details

Graham V. Weinberg

Address all correspondence to: Graham.Weinberg@defence.gov.au

National Security, Intelligence, Surveillance, Reconnaissance Division, Defence Science, Technology Group, Edinburgh, South Australia, Australia

References

[1] Minkler, G., Minkler, J.: CFAR: The Principles of Automatic Radar Detection in Clutter, Magellan, Baltimore, 1990.

[2] Finn, H. M., Johnson, R. S.: Adaptive Detection Model with Threshold Control as a Function of Spatially Sampled Clutter-Level Estimates, *RCA Review*, 1968, **29**, pp. 414–464.

[3] Nitzberg, R.: Low-Loss Almost Constant False-Alarm Rate Processors, *IEEE Transactions on Aerospace and Electronic Systems*, 1979, **AES-15**, pp. 719–723.

[4] Weiss, M.: Analysis of Some Modified Cell-Averaging CFAR Processors in Multiple-Target Situations, *IEEE Transactions on Aerospace and Electronic Systems*, 1982, **AES-18**, pp. 102–114.

[5] Rohling, H.: Radar CFAR Thresholding in Clutter and Multiple Target Situations, *IEEE Transactions on Aerospace and Electronic Systems*, 1983, **AES-19**, pp. 608–621

[6] Nitzberg, R.: Clutter Map CFAR Analysis, *IEEE Transactions on Aerospace and Electronic Systems*, 1986, **AES-22**, pp. 419–421.

[7] Gandhi, P. P., Kassam, S. A.: Analysis of CFAR Processors in Nonhomogeneous Background, *IEEE Transactions on Aerospace and Electronic Systems*, 1988, **24**, pp. 427–445.

[8] Chen, W.-S., Reed, I. S.: A New CFAR Detection Test for Radar, *Digital Signal Processing*, 1991, **1**, pp. 198–214.

[9] Al-Hussaini, E. K., El-Mashade, M. B.: Performance of Cell-Averaging and Order-Statistic CFAR Detectors Processing Correlated Sweeps for Multiple Interfering Targets, *Signal Processing*, 1996, **49**, pp. 111–118.

[10] Hamadouche, M, Barakat, M., Khodja, M.: Analysis of the Cutter Map CFAR in Weibull Clutter, *Signal Processing*, 2000, **80**, pp. 117–123.

[11] Laroussi, T., Barkat, M: Performance Analysis of Order-Statistic CFAR Detectors in Time Diversity Systems for Partially Correlated Chi-Square Targets and Multiple Target Situations: A Comparison, *Signal Processing*, 2006, **86**, pp. 1617–1631.

[12] Erfanian, S., Vakili, V. T.: Introducing Excision Switching-CFAR in K Distributed Sea Clutter, *Signal Processing*, 2009, **89**, pp. 1023–1031.

[13] Zhang, R., Sheng, W., Ma, X.: Improved Switching CFAR Detector for Non-Homogeneous Environments, *Signal Processing*, 2013, **93**, pp. 35–48.

[14] Zhang, R.-I., Sheng, W.-X., Ma, X.-F., Han, Y.-B.: Constant False Alarm Rate Detector based on the Maximal Reference Cell, *Digital Signal Processing*, 2013, **23**, pp. 1974–1988.

[15] Zaimbashi, A.: An Adaptive Cell Averaging-Based CFAR Detector for Interfering Targets and Clutter-Edge Situations, *Digital Signal Processing*, 2014, **31**, pp. 59–68.

[16] Balleri, A., Nehorai, A., Wang, J.: Maximum Likelihood Estimation for Compound-Gaussian Clutter with Inverse-Gamma Texture, *IEEE Transactions on Aerospace and Electronic Systems*, 2007, **43**, pp. 775–779.

[17] Farshchian, M., Posner. F. L.: The Pareto Distribution for Low Grazing Angle and High Resolution X-Band Sea Clutter, *IEEE Radar Conference*, 2010, pp. 789–793.

[18] Weinberg, G. V.: Assessing the Pareto Fit to High Resolution High Grazing Angle Sea Clutter, *IET Electronics Letters,* 2011, **47**, pp. 516–517.

[19] Sangston, K. J., Gini, F., Greco, M. S.: Coherent Radar Target Detection in Heavy-Tailed Compound Gaussian Clutter, *IEEE Transactions on Aerospace and Electronic Systems,* 2012, **48**, pp. 64–77.

[20] Weinberg, G. V.: Constant False Alarm Rate Detectors for Pareto Clutter Models, *IET Radar, Sonar and Navigation*, 2013, **7**, pp. 153–163.

[21] Weinberg, G. V.: General Transformation Approach for Constant False Alarm Rate Detector Development, *Digital Signal Processing,* 2014, **30**, pp. 15–26.

[22] Weinberg, G. V.: Constant False Alarm Rate Detection in Pareto Distributed Clutter: Further Results and Optimality Issues, *Contemporary Engineering Sciences,* 2014, **7**, pp. 231–261.

[23] Weinberg, G. V.: Management of Interference in Pareto CFAR Processes using Adaptive Test Cell Analysis, *Signal Processing,* 2014, **104**, pp. 264–273

[24] Weinberg, G.V.: Development of an Improved Minimum Order Statistic Detection Process for Pareto Distributed Clutter, *IET Radar, Sonar and Navigation,* 2015, **9**, pp. 19–30.

[25] Weinberg, G.V.: Examination of Classical Detection Schemes for Targets in Pareto Distributed Clutter: Do Classical CFAR Detectors Exist, as in the Gaussian Case?, *Multidimensional Systems and Signal Processing,* 2015, **26**, pp. 599–617.

[26] Beaumont, G. P.: Intermediate Mathematical Statistics, Chapman and Hall, London, 1980.

[27] Watts, S.: Radar Detection Prediction in Sea Clutter using the Compound K-Distribution Model, *IEE Proceedings, Part F,* 1985, **132**, pp. 613–620.

[28] Stacy, N. J. S., Burgess, M. P., Muller, M. R., Smith, R.: Ingara: An Integrated Airborne Imaging Radar System, *Proceedings of the International Geoscience and Remote Sensing Symposium*, 1996, pp. 1618–1620.

[29] Stacy, N., Crisp, D., Goh, A., Badger, D., Preiss, M.: Polarimetric Analysis of Fine Resolution X-Band Sea Clutter Data, *Proceedings of the International Geoscience and Remote Sensing Symposium,* 2005, pp. 2787–2790.

[30] Meng, X. W.: Performance Analysis of OS-CFAR with Binary Integration for Weibull Background, *IEEE Transactions on Aerospace and Electronic Systems,,* 2013, **49**, pp. 1357–1366.

[31] Weinberg, G. V., Kyprianou, R.: Optimised Binary Integration with Order Statistic CFAR in Pareto Distributed Clutter, *Digital Signal Processing*, 2015, **42**, pp. 50–60.

[32] Frey, T. L.: An Approximation for the Optimum Binary Integration Threshold for Swerling II Targets, *IEEE Transactions on Aerospace and Electronic Systems*, 1996, **32**, pp. 1181–1184.

10

Validation of Instrument Measuring Continuous Variable in Medicine

Rafdzah Zaki

Abstract

In medicine, accurate measurement of clinical values is vital, either at the stage of health screening, diagnosing cases or making prognosis. There are numerous instruments or machines that have been invented for the purpose of measuring various clinical variables such as blood pressure, glucose level, body temperature and oxygen level. When a new method of measurement or instrument is invented, the quality of the instrument has to be assessed. This chapter will focus on the application of statistical methods used to analyse continuous data in a method comparison study or validation study in medicine. The concept of validity and analysis in method comparison study will be discussed. This chapter also reviews the theoretical aspects of several common methods and approaches that have been used to measure agreement and reliability including the Bland-Altman limits of agreement (LoA) and intra-class correlation coefficient (ICC). Issues related to method comparison studies will be highlighted, which include the evaluation of agreement and reliability in a single study, the application of multiple statistical methods and the use of inappropriate methods in testing agreement and reliability. Finally, the importance of education in method comparison studies among medical professional will be emphasized.

Keywords: agreement, reliability, medical instrument, continuous variable

1. Introduction

In medicine, accurate measurement of clinical values is vital, either at the stage of health screening, diagnosing cases or making prognosis. For example, accurate measurement of blood pressure, heart rate and oxygen level is crucial for monitoring patients under general anaesthesia in surgery. Inaccurate measurement of these variables will result in inappropriate management of the patient, thus putting the patient's life at risk.

Most of the important variables measured in medicine are in numerical forms or continuous in nature, such as blood pressure, glucose level, oxygen level, weight, height, body temperature, creatinine level, albumin level and many other clinical values. There are numerous instruments or machines that have been invented for the purpose of measuring various variables. Some measurements are obtained by using invasive techniques and expensive procedures. Consequently, new instruments and tests are constantly being developed and fashioned to provide complementary meaningful information to the search for information, with the aim of providing cheaper, non-invasive, more convenient and safe methods. Whether a test's outcome can provide trustworthy judgements or decisions depend particularly on the measurement quality of the test [1].

When a new method of measurement or instrument is invented, the quality of the instrument has to be assessed. We want to know by how much the value of measurements obtained using new method differs from the old method, or from the gold standard. Information provided by any clinical instrument cannot be trusted and licitly used in any judgement and decision-making process if the measurement quality has not been evaluated. This is where a method comparison study or a validation study comes into medicine. Clinimetric properties indicating that the test is reliable and valid should be considered as fundamental for determining the measurement quality of any test [2]. In general, clinimetric refers to the development of methodological and statistical methods applicable in clinical medicine in order to assign numbers or scores to observable clinical events [3, 4].

1.1. Validity

An instrument is considered to be valid if it measures what it is intended to measure [3]. The term 'validity' actually has a wide range of classification and definition. In clinical research, current and accepted validity concepts include *criterion* validity, *construct* validity and *content* validity, the first two being the most relevant for performance-based tests [5].

Criterion validity is used to examine the extent to which a measurement instrument provides the same results as the gold standard [5]. This type of validity is the most powerful in terms of its usefulness, and is divided into two types: *concurrent* validity and *predictive* validity. Of these, *concurrent validity* is the most used method. This is when we are trying to compare a new measurement tool with the criterion measure, both of which are given at the same time [5, 6]. The new tool is usually simpler, cheaper or less invasive compared to the standard or currently used tools. In contrast, in *predictive validity*, the criterion will not be available until sometime in the future. When no gold standard is available, the common alternative is to use an accepted and well-grounded reference test to relate to the evaluated test [7, 8]. Generally, this form of validity is used in developing instruments that allow us to get earlier answers or to give earlier predictions than current instruments can provide [5].

Construct validity refers to the degree to which a test measures a hypothetical, non-observable construct and this validity can be established by relating the test to outcomes of other instruments [1, 5]. It is used when we dealing with more abstract variables or factors that cannot be measured directly for example level of anxiety and pain [5]. We cannot see or directly measure anxiety, but we can observe other factors related to anxiety (according to theory) such as sweaty palm and tachycardia. The proposed underlying factors are referred to as *hypothetical construct* or simply known as *constructs* [5]. So, *construct validity* is the next best option in the absence of an acceptable gold standard. The measurement of instrument under study will be compared with other instruments that claim to measure the same construct [5].

Content validity is a closely related concept, consisting of a judgement whether the instrument samples all the relevant or important content or domains [5, 9]. Content validity can be claimed when a test logically and obviously measures what it purposes to measure [5, 6]. The relationship between the phenomenon being measured and the test score(s) is determined by a panel of experts or researchers [6].

1.2. Reproducibility

Another approach in assessing the quality of measuring instrument is to assess the *reproducibility* of the instrument. This is when we are interested to know whether the new instrument is able to produce similar values as that predicted by the old or standard instrument. In the literature, terms reproducibility is often used interchangeably with the reliability, repeatability, consistency, agreement and stability [9]. Recently, de Vet et al. [3, 10] advocated that reproducibility is the proper term to use in clinical research, making the distinction between two aspects that are important for clinical interpretation: reliability and agreement.

1.2.1. Agreement

Agreement assesses how close the results of repeated measurements are to the 'true value' or the criterion value [10]. So, agreement actually concerns accuracy or validity; more specifically, concurrent validity. An instrument with good agreement will be able to produce accurate repeated measurements in the same person [10]. Thus, agreement parameters are important in instruments that are used for evaluative purposes. In evaluative measurement instruments, the variability between individuals in a population is not important, in comparison to the variability within an individual [10]. This is because, in some clinical settings, we want to detect differences or changes within the same individual, and not how much difference is the individual's value compared to another person's, or with the population. For example, in antenatal clinics we are interested in the weight gain of a mother throughout her pregnancy, and not how much her weight differs from the others'.

Agreement parameters estimate the *measurement error* in repeated measurements. When the measurement error is large, small changes cannot be distinguished from the measurement error [10]. The smaller the measurement error, the smaller the changes that can be detected beyond the measurement error, and the more appropriate the instrument is for evaluative purposes. Thus, for an instrument to be used to evaluate changes over time, such as changes in blood pressure after receiving antihypertensive therapy, it is important for us to ensure the agreement or the accuracy of the instrument.

1.2.2. Reliability

Reliability measures the extent to which the test results can be replicated [11]. For example, if we measure body weight using a scale five times, ideally all five measurements should be the same.

Reliability is concerned with precision. It also represents the extent to which individuals can be distinguished from each other, despite the variability of repeated measurements in one person or subject (i.e. measurement error) [10]. In contrast with agreement, reliability measures the variability between people or subjects. This measurement tells us how well the measured value in one person can be distinguished from another [10]. Thus, reliability parameters are important when measurement instruments are used for discriminative purposes; for example, to decide whether a certain value is normal or abnormal, and when the measurement from the instrument is involved in important decisions, such as whether treatment is required or not.

In clinical practice, the cut-off for normal and abnormal values is usually well established by clinical guidelines, which are produced based on extensive reviews of available evidence. Reliable instruments should be able to provide values that will allow doctors or clinicians to distinguish whether their patients are in the normal or abnormal group. For instance, if we take the blood pressure of one patient five times, all the values should be almost the same, and the values should give us an idea whether the patient's blood pressure is normal or not.

An acceptable range of reliability will vary depending on the circumstances [5]. For example, if repeated measurements of a weighing scale are found to vary around the 'true' weight by 0.5 kg, the reliability of this weighing scale would be acceptable if the measurements are only to be done on an adult population, but not reliable when used to weigh new-born babies in the hospital. This is because differences of 0.5 kg in weight in an adult represent only a very small percentage of an adult body weight, and will not affect him or her clinically. In contrast, a difference of 0.5 kg represents a large proportion of body weight for a new-born baby.

1.3. Agreement versus reliability

Although the terms 'agreement' and 'reliability' carry different meanings, they are sometime used interchangeably in medical literature. To illustrate the concept of agreement and reliability in more simple language, imagine if we have three target boards (see **Figure 1**) that show the results of five repeated measurements of body weight of the same person, using three different scales (A, B and C). The centre of each board indicates the true value. **Figure 1(A)** shows that after taking five measurements using scale A, the results of the measurements are scattered all over the target board. This suggests that the measurements are not near each other (poor reliability), and are not near their intended target or true value (poor agreement).

Figure 1(B) shows that all the five measurements from scale B appear in more or less the same location on the target board, but not in the centre of the target board. This suggests that five different measurements were almost the same (good reliability), but they did not hit the intended target (poor agreement). **Figure 1(C)** shows that all the five measurements from scale C are close to each other (good reliability), and hit the centre of the target board (good agreement).

In most clinical situations, we use the same instrument to evaluate changes over time and also to differentiate values from the normal or abnormal cut-off point (which is usually derived from population-based studies). One of the examples of this situation is in the screening of hypertension cases, and the assessment of the reduction of blood pressure post-treatment, in a clinic or health centre. Both blood pressure measurements are performed using the same blood pressure machine, or sphygmomanometer. So, agreement and reliability parameters are equally important in determining the quality of instruments. In fact, it is difficult to be certain about the agreement of an instrument if the instrument is not reliable. Similarly, a precise instrument or instrument with good reliability will not necessarily measure the 'true' value. Therefore, when comparing two instruments, or methods of measurement, we should consider assessing the *repeatability* of the instrument, which covers both agreement (accuracy) and reliability (precision).

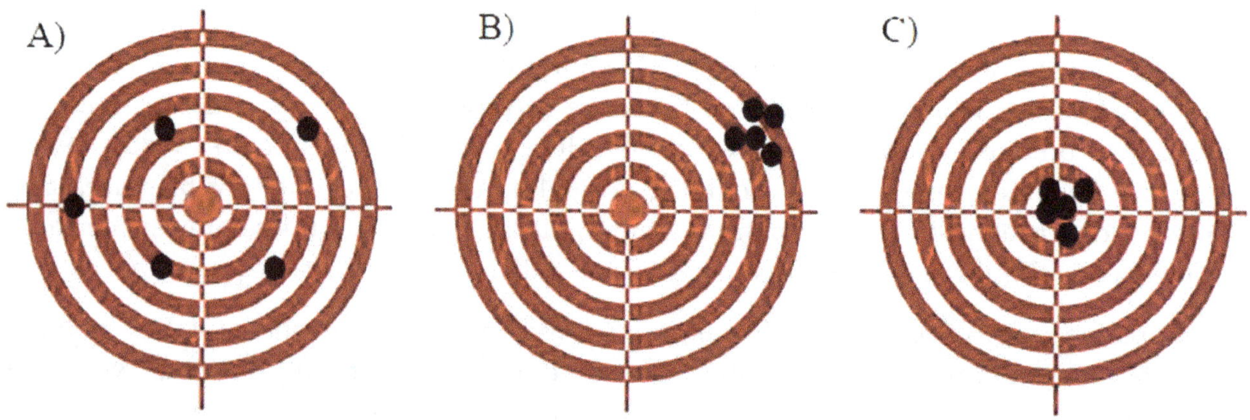

Figure 1. Results of measurements of body weight using three different scales A, B and C.

2. Issues related to method comparison studies

2.1. Evaluation of agreement and reliability in a single study

Agreement and reliability are both important in assessing the quality of instruments. An instrument with high agreement will not be useful if it is unreliable. Ideally, these parameters should be assessed together. However, recent systematic reviews showed that this is not commonly followed in practice, especially with respect to agreement studies [12, 13]. Most (71%) of the reliability studies, measured agreement at the same time [13]. However, only 30% of agreement studies found assessed reliability [12]. Researchers tend to focus on one aspect of quality when validating instruments, although there is a possibility of agreement and reliability studies being conducted separately for the same instrument. Nonetheless, it is important to ensure the reliability of the instrument first, before testing for agreement, because it is impossible to assess the agreement of an unreliable instrument.

2.2. Inappropriate application of statistical method

Thousands of validation studies have been conducted in the past. Various statistical tests have been used to test for agreement and reliability [14–16]. Some of the methods that were used were inappropriate. Correlation coefficient (r), coefficient of determination (r^2), regression coefficient and means comparison have been shown to be inappropriate for the analysis in method comparison study. This has been discussed by Altman and Bland, since the 1980s [15] and also by Daly and Bourke [17]. Reasons for why those methods are inappropriate for the analysis in method comparison study will be discussed in Sections 3 and 4.

One example of the inappropriate application of statistical methods in method comparison study is in the study to explore the suitability of existing formulas to estimate the body surface area (BSA) of new-borns [18]. The authors compared different methods of estimation of body surface area in newborn and used correlation coefficient to determine the agreement of those methods [18]. In one of their results, the authors described that the method of estimating body surface area (BSA) using the BSA-Meban was most similar to the BSA-Mean, by having a mathematically perfect correlation with $r = 1.00$ ($p = 0.000$) [18]. However, their conclusion was obviously inappropriate because the correlation coefficient only measures linear relationship, and does not suggest that the two methods give similar results.

Another example of the inappropriate application of the Pearson correlation coefficient was demonstrated in one study conducted in Greece [19]. The authors aimed to assess the validity of a new motorised isometric dynamometer for measuring strength characteristics of elbow flexor muscles. They set the criteria of the Pearson correlation coefficient's (r) values >0.97 to demonstrate that high agreement occurred between measures, and with $r = 0.986$, they concluded that the new dynamometer was accurate [19].

The use of inappropriate methods for the assessment of agreement and reliability will, undoubtedly, result in an inappropriate interpretation of the results and conclusions on the quality of an instrument. Consequently, this might result in the application of invalid equipment in medical practice, and will jeopardise the quality of care given to patients. The proportion of studies with inappropriate statistical methods might reflect the proportion of medical instruments that have been validated using inappropriate methods in current clinical practice.

As found in recent systematic reviews, 19% of reliability studies [13] and 10% of agreement studies [12] used inappropriate methods, which means that there is a distinct possibility that some medical instruments or equipment used currently were validated using inappropriate methods, with consequently erroneous conclusions being drawn from these methods. This equipment, therefore, may not be as precise or accurate as believed, which could, potentially, affect the management of patients, the quality of care given to patients and, worse, it could cost lives.

Altman and Bland [15] proposed a method for agreement analysis in their original 1983 article. Later, they drew the attention of the medical professionals to this area in an article in The Lancet [20]. Their article [20] received very high citation [21]. The popularity of the Bland-Altman method was thought, owing to its simplicity, practicality and ability to detect bias, when compared to other methods [16].

The issue of which method is the best is still debatable, and almost all methods have been criticised, especially for the agreement study. Even the Bland-Altman method has been criticised. Hopkins [22] demonstrated that the Bland-Altman plot indicates, incorrectly, that there is systematic bias in the relationship between two measures. Recent study also showed that there is overestimation of bias in the Bland-Altman analysis [23].

2.3. Application of multiple methods

According to recent systematic reviews conducted, most reliability studies (86%) relied on a single statistical method to assess reliability [13], in contrast with agreement studies where most of the studies (65%) used a combination of statistical methods [12]. The application of multiple or a combination of methods, particularly in the assessment of agreement, suggests that there is no consensus among researchers on which method is the best statistical method for measuring agreement. One example of the multiple application of method is in one study that testing the accuracy of peak flow meters [24]. In this study, the authors applied three statistical methods (Pearson's correlation coefficient, comparing mean (significant test) and the Bland-Altman method) to assess for agreement of peak flow meters [24].

A strong reason for using multiple methods in assessing agreement and reliability is that each statistical method has its strengths and weaknesses. The usage of multiple methods in method comparison studies has the advantage of compensating for the limitations of any one single method [14, 25]. As long as the methods chosen are appropriate for it purposes. However, another possible reason for using multiple methods is the researcher's limited understanding of the statistical methods for agreement and reliability. This is probably the reason for the application of multiple inappropriate statistical methods in a single study; for example, the use of both correlation coefficient and significance test of the difference between means, to test for agreement and reliability. Both of these methods have been clearly shown to be inappropriate statistical methods to assess agreement and reliability [15, 17].

3. Most commonly used methods to assess agreement

In 2012, Zaki et al. [12] review the statistical methods used to measure the agreement of equipment measuring continuous variables in the medical literature. The most common method to assess agreement is the Bland-Altman limits of agreement (LoA), followed by correlation coefficient (r), comparing means, comparing slope and intercept and intraclass correlation coefficient. However, some of these methods were inappropriate to assess agreement.

3.1. Bland-Altman limits of agreement

Bland-Altman limits of agreement were found to be the most commonly used method to assess agreement in medical literature. In 1983, Bland and Altman introduced limits of agreement (LoA) to quantify agreement [15]. They proposed a scatter plot of the differences of two measurements against the average of the two measurements, and later it becomes a graphical presentation of agreement (see **Figure 2**). Bland and Altman [20] stated that it is very unlikely for two different methods or instruments to be exactly in agreement, or give identical results for all individuals. However, what is important is how close the values obtained by the new method (predicted values) are to the gold standard method (actual values). This is because a very small difference in the predicted and the actual value will not have an effect on decisions of patient management [20]. So they started with an estimation of the difference between measurements by two methods or instruments [20]. To construct limits of agreement, first we need to calculate the mean and standard deviation of these differences. The formula for limits of agreement (LoA) is given as [20]:

$$\text{LoA} = \text{mean difference} \pm 1.96 \times (\text{standard deviation of differences}) \quad (1)$$

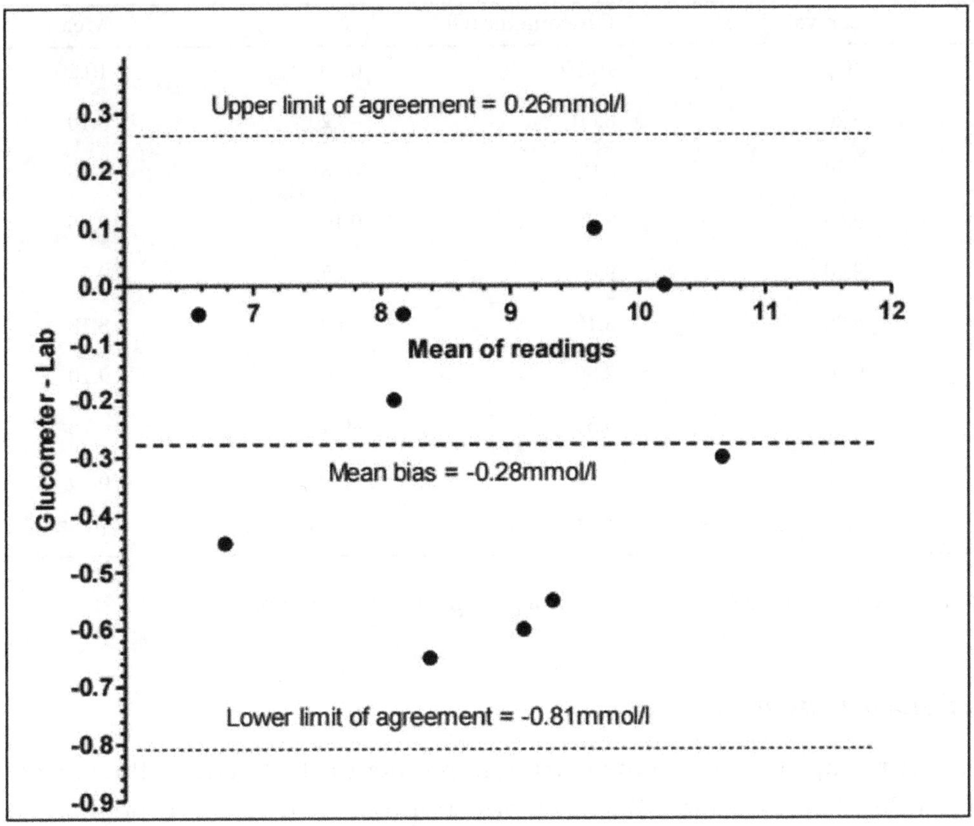

Figure 2. The Bland-Altman plot.

So, 95% of differences should lie within these limits. To illustrate this, we can use the data from **Table 1** (adapted from Table 12.5, interpretation and uses of medical statistics) [17], which compared the values from the glucometer and laboratory. If we apply the data from **Table 1**, the first step of the analysis is to calculate the difference and mean. The mean difference for the data is −0.28 mmol/l, and the standard deviation of difference is 0.27 mmol/l. This makes the LoA = −0.81 to 0.26 mmol/l.

Limits of agreement give us the range of how much one method is likely to differ from another. So it is all about the differences. If we are testing a new method B against the old method A, and the difference is calculated from A–B, then a positive value of limits of agreement means A > B, or new method B underestimates the new method A. If a negative value of limits of agreement means A < B, or the new method B overestimates the old method A. So, the result of Bland-Altman analysis between glucometer and laboratory values (**Table 1**) can be shown as follows:

$$\begin{aligned} \text{Differences} &= \text{Glucometer–Laboratory} \\ \text{Mean difference} &= -0.28 \text{ mmol/l} \\ \text{Limits of Agreement} &= -0.81\text{–}0.26 \text{ mmol/l} \end{aligned} \tag{2}$$

This means that, on average, the glucometer measures 0.28 mmol/l less than the laboratory. Also, 95% of the time the glucometer reading will be somewhere between 0.81 mmol/l below and 0.26 mmol/l above the laboratory values.

Patient	Lab value (*L*)	Glucometer (*G*)	*G–L*	Mean
1	10.20	10.20	0.00	10.20
2	8.20	8.00	–0.20	8.10
3	8.70	8.05	–0.65	8.38
4	9.60	9.70	0.10	9.65
5	9.60	9.05	–0.55	9.33
6	8.20	8.15	–0.05	8.18
7	9.40	8.80	–0.60	9.10
8	7.00	6.55	–0.45	6.78
9	6.60	6.55	–0.05	6.58
10	10.80	10.50	–0.30	10.65

Table 1. Hypothetical data of blood glucose level from a glucometer and laboratory.

3.2. Correlation coefficient

One of the favourite approaches in measuring agreement is to calculate the correlation coefficient (r) [11, 15, 26]. This method is the next most popular method after the Bland and Altman method [12]. The first approach in this analysis is to make a scatter plot, and then to calculate product-moment correlation coefficient [11]. To calculate the product moment correlation coefficient (r), variables for each pair of measurements are labelled as X and Y. The formula for the correlation coefficient r is given as:

$$r = \frac{\sum XY - \frac{\sum X \sum Y}{N}}{\sqrt{\left(\sum X^2 - \frac{(\sum X)^2}{N}\right)\left(\sum Y^2 - \frac{(\sum Y)^2}{N}\right)}} \quad (3)$$

If we use an example from data presented in **Table 1** (to compare blood sugar levels from glucometer and laboratory values), the Pearson correlation coefficient (r) is 0.9798 with a 95% confidence interval of 0.9139–0.9954, and p-value <0.0001 (analysis using SPSS 17.0 software). The null hypothesis here is that the measurements of blood glucose level by the two methods (glucometer and laboratory) are not related linearly. With a very small p-value, we can reject this null hypothesis and propose an alternative hypothesis: there is a linear relationship between the measurements of glucose level by the two methods (glucometer and laboratory). Some people will interpret this as being that there is an agreement between the two instruments. This is another mistake conducted by many researchers [15].

Correlation is a measure of association, and only measures the strength of linear relationship [11]. Strong correlation does not mean strong agreement. To demonstrate the inappropriate use of correlation, let's double the value of glucometer from **Table 1** so that it is obvious that there is no agreement between the glucometer and the laboratory value (see **Table 2**). Despite this, the correlation analysis of data from **Table 2** will give exactly the same Pearson correlation coefficient (r) of 0.9798, with a similar 95% confidence interval (CI) of 0.9139–0.9954.

Of course, the two instruments (glucometer and laboratory measurement) do not agree, but the correlation coefficient value is still very high, suggesting a strong correlation or association.

Agreement is assessing a different aspect of relationship between two measurements as compared to the correlation coefficient. The correlation coefficient reflects the noises and the direction of a linear relationship [27, 28]. Perfect correlation occurs if all the points lie along any straight line (see **Figure 3**), and so data with poor agreement can produce a high or strong association [20, 29]. Furthermore, data covering an extensive (wide) range of values will appear to be more highly correlated than if it covers a narrow range [20]. Therefore, it is clear that correlation is not an appropriate method for testing agreement.

Lab value	Glucometer	Glucometer ×2
10.20	10.20	20.40
8.20	8.00	16.00
8.70	8.05	16.10
9.60	9.70	19.40
9.60	9.05	18.10
8.20	8.15	16.30
9.40	8.80	17.60
7.00	6.55	13.10
6.60	6.55	13.10
10.80	10.50	21.00

Table 2. Hypothetical data of blood glucose value.

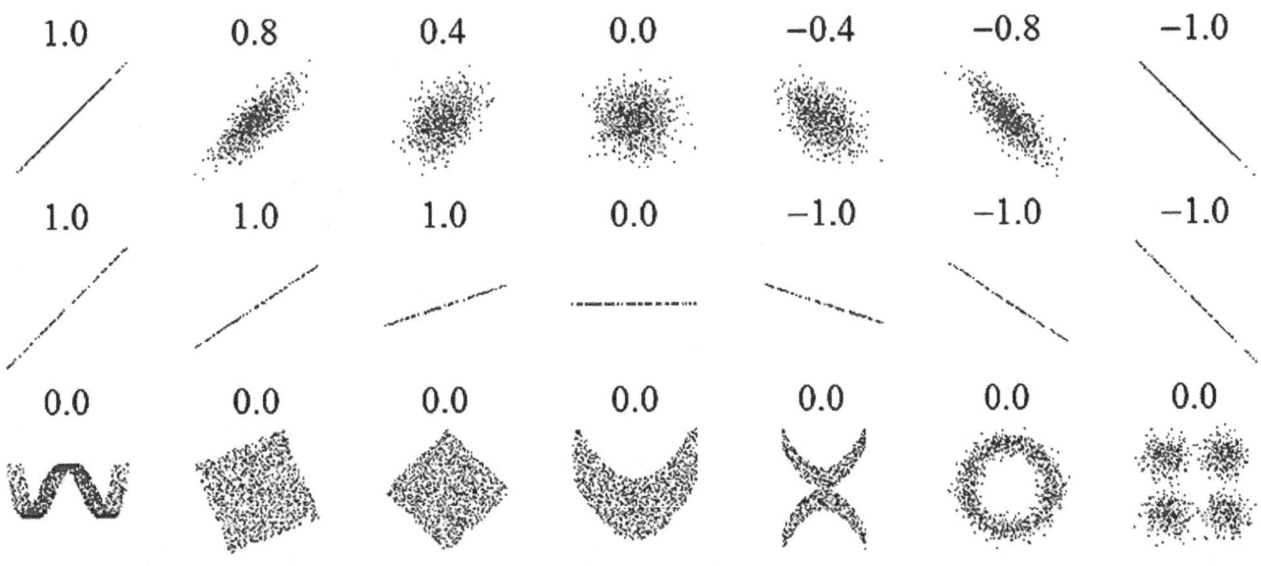

Figure 3. Correlation coefficient values, the noises and direction of a linear relationship [29].

Some people use the *coefficient of determination* (r^2) parameter as a measure of agreement. One example of the application of this method is in a recent study [30] on the accuracy evaluation of point-of-care glucose analysers in the Saudi Arabian market. The authors compare the blood glucose readings from five different types of glucose analysers with the results from laboratory analysis. Their aim was to test the accuracy of the devices. In one of their results, the authors described that the Nova StatStrip device showed an excellent performance that almost agreed and correlated perfectly with the lab results, because the $r^2 = 0.99$ [30]. However, the use of the *coefficient of determination* (r^2) is inappropriate because r^2 is obtained from correlation coefficient r, which is a wrong method to measure agreement. *Coefficient of determination* (r^2) is used to state the proportion of variance in the dependent variables that is explained by the regression equation or model [17]. The more closely the points are dispersed around the regression line in the scatter plot, the higher the proportion of variation explained by the regression line, thus the greater the value of r^2 [11]. So it applies a similar concept to the correlation coefficient.

3.3. Comparing means

The third most popular method used is comparing means of readings from two instruments [12]. In this method, the means of readings from two instruments are compared. The test of significance is then carried out to test the null hypothesis that there is no difference between the means of readings from the two instruments.

In assessing agreement, the same measurement of similar subjects will be taken using different instruments. Therefore, a paired *t*-test is usually used to test the hypothesis. Here, we want to know whether the difference observed is the true difference or has only occurred by chance when there was really no difference in the population. If the difference is truly occurring, and the null hypothesis is not true, then the alternative hypothesis must be true. So, in this case, the alternative hypothesis is that there is a significant difference between the mean of reading from the two instruments.

People have interpreted non-significance results to mean that there is not enough evidence to show that the two means differ (i.e. no differences), thus there is an agreement between the two groups, and vice versa. An example of this inappropriate approach is in a study conducted in Sweden on the assessment of left ventricular volumes, using simplified 3-D echocardiography and computed tomography [31]. However, the paired t-test with non-significant results does not indicate agreement. The reason for this is that the value of mean is affected by the value of each piece of data, especially when there is an outlier. Distribution of differences between the instruments can lead to a difference in means being non-significant. It is possible that poor agreement between the two instruments can be hidden in the distribution of differences, and thus the two methods can appear to agree [17]. To illustrate this example, we have a hypothetical dataset comparing the measurements from standard instrument A, with the new instruments B and C (**Table 3**).

From the dataset (**Table 3**), it is obvious that the two new instruments (B and C) do not agree with the standard instrument A. The mean and standard deviation for the three groups are all

Patient	A	B	C
1	1	1	5
2	2	3	4
3	3	2	3
4	4	5	2
5	5	4	1

Table 3. Hypothetical dataset for instruments A, B and C.

the same: the mean is equal to 3.0 and standard deviation is equal to 1.58. If we compare the readings from instruments A and B, using a paired *t*-test, the results will be:

$$\begin{aligned}&\text{Mean difference (confidence interval)} = 0\,(-1.24\text{–}1.24)\\&\text{Standard deviation of differences} = 1.0\\&p\text{-value} = 1.0\end{aligned} \tag{4}$$

So, from this analysis, we can conclude that there is no difference between the mean reading of instruments A and B. If we are saying that non-significant results indicate an agreement, this suggests that there is an agreement between instruments A and B. However, we know that this is not true. Similarly, the result will be not significant when we compare the mean reading of instruments A and C, where the results will be:

$$\begin{aligned}&\text{Mean difference (confidence interval)} = 0\,(-3.93\text{–}3.93)\\&\text{Standard deviation of differences} = 3.16\\&p\text{-value} = 1.0\end{aligned} \tag{5}$$

Again, this does not suggest that there is an agreement between instruments A and C. The inappropriate application of the test of significance, as a test for agreement, has also been discussed earlier in the article by Altman and Bland [15]. What matters in agreement is that each reading from the standard instrument should be repeated by the second instrument. We are not interested in the mean of readings by each instrument, but are interested in each individual reading. Therefore, comparing means using a significance test is not an appropriate method for assessing agreement.

3.4. Intra-class correlation coefficient

The intra-class correlation coefficient (ICC)is another popular method found used to test for agreement [12]. ICC was devised initially to assess the relationship between variables within classes, or reliability. However, it was then used to assess agreement, to avoid the problem of linear relationship being mistaken for agreement in the product moment correlation coefficient (r) [26, 32]. Different assignments of measurements of X and Y, in the calculation of the correlation coefficient (r), would produce different values of r. To overcome some of the limitations of the correlation coefficient (r), the ICC averages the correlations among all the possible ordering of the pairs [33]. The ICC also extends to more than two observations, in contrast with the correlation coefficient (r) [11]. In general, the ICC is a ratio of two variances:

$$\text{ICC} = \frac{\text{Variance owing to rated subjects}}{\text{Variance owing to subjects + Eerror}} \quad (6)$$

The value of the ICC can theoretically vary from 0 to 1, where 0 indicates no reliability or disagreement in the agreement study. The ICC of one indicates perfect reliability, or perfect agreement. There are different types of ICC that have been described by Shrout and Fleiss [33]. McGraw and Wong [35] expanded the Shrout and Fleiss system to include two more general forms of ICC. Weir [34] summarised different types of ICC, based on models introduced by Shrout and Fleiss [33] and McGraw and Wong [35] (see **Table 4**).

Shrout and Fleiss [33]	Computational formula	McGraw and Wong [35]	Model
1, 1	$\frac{MS_B - MS_W}{MS_B + (k-1)MS_W}$	1	One-way random
1, *k*	$\frac{MS_B - MS_W}{MS_B}$	*K*	One-way random
	Use 3, 1	*C*, 1	Two-way random
	Use 3, *k*	*C*, *k*	Two-way random
2, 1	$\frac{MS_S - MS_E}{MS_S + (k-1)MS_E + (k(MS_T - MS_E)/n)}$	*A*, 1	Two-way random
2, *k*	$\frac{MS_S - MS_E}{MS_S + (k(MS_T - MS_E)/n)}$	*A*, *k*	Two-way random
3, 1	$\frac{MS_S - MS_E}{MS_S + (k-1)MS_E}$	*C*, 1	Two-way fixed
3, *k*	$\frac{MS_S - MS_E}{MS_S}$	*C*, *k*	Two-way fixed
	Use 2, 1	*A*, 1	Two-way fixed
	Use 2, *k*	*A*, *k*	Two-way fixed

MS_B, Between-subjects mean square; MS_E, error mean square; MS_S, subjects mean square; MS_T, trials mean square; MS_W, within-subjects mean square.

Table 4. Different types of ICC.

Shrout and Fleiss [33] suggested three main models: Model 1 is a one-way fixed model; Model 2 is a two-way random model; and Model 3 is a two-way fixed model. The model is represented in the format of ICC (a, b). The value of '*a*' can be 1, 2 or 3 (this depends on the three main models). For value '*b*', when $b = 1$, this suggests single measures ICC and $b = k$ suggests averaged measures ICC [34].

In the ICC model suggested by McGraw and Wong [35], the designation '*C*' refers to consistency and '*A*' refers to absolute agreement. The '*A*' model considers both fixed and systematic error, whereas the '*C*' model only considers fixed error [34, 35].

Although a total of 10 ICC models were summarised by Weir [34], there are similarities in some of the ICC formula for different types of ICC. This is because the difference between the random model and the fixed model is not in the calculation but in the interpretation of the ICC [35]. According to Shrout and Fleiss [33], there is only one ICC that measures the extent of absolute agreement, and that is ICC (2, 1), which is based on the two-way random-effects ANOVA (analysis of variances) [14, 33]. This model is similar to ICC (*A*, 1), as suggested by McGraw and Wong [35].

The ICC (*C*, 1) for consistency simply compares the consistency between trials. For example, for the hypothetical data from **Table 5**, will produce ICC (*C*, 1) = 1.0, which is interpreted as a perfect agreement. However, the absolute agreement ICC, or ICC (*A*, 1), compares both the consistency between trails and the agreement between ratings. So, the same pairs of data from **Table 5** will produce ICC (*A*, 1) = 0.67, which suggests some degree of disagreement (or moderate agreement).

Patient	First reading	Second reading
1	2	4
2	4	6
3	6	8

Table 5. Hypothetical dataset of repeated measurements from instrument A.

However, the use of ICC in assessing agreement has been criticised by Bland and Altman [32]. In testing the agreement of instruments, the new method will usually be compared to the standard instrument [32]. The aim of testing is to ensure that the new method will produce the same measurements as the standard instrument (i.e. good agreement). This can also mean that the new method is designed to provide similar predictions of measurement as the standard instrument. So, there is clear ordering of the two variables, where the measurements from the standard instrument are usually denoted as X and measurements of the new method are denoted as Y.

The ICC also ignores the ordering and treats both methods as a random sample from a population of methods [32]. In an agreement study, there are two specific methods that will be compared, not two instruments chosen at random from some population. Another assumption in the ICC model, which is quite unjustified in methods comparison study, is that the measurement error of both methods has to be the same [32]. The main purpose in testing agreement is to identify the measurement error of the new instrument in comparison to the standard instrument. Another issue with ICC is that it is influenced by the range of data. If the variance between subjects is high, the reliability will certainly appear to be high [14].

3.5. Comparing slopes and y-intercepts

Often, in testing for agreement, the slope is tested against one. The argument is that if the two methods or instrument are equivalent (i.e. if it measures the same variable of the same subject, both instruments will give the same reading), thus the slope of the straight line will be one [15].

Straight line equation will show the relationship between two variables, and can be expressed as: $y = \alpha + \beta x$, where $'y'$ is the predicted or expected value for any given value of $'x'$, while $'\alpha'$ is the intercept of the straight line with the y-axis and $'\beta'$ is the slope. The values of both $'\alpha'$ and $'\beta'$ are constant. The slope $'\beta'$ is also called the *regression coefficient*, and measures the amount of change in the $'y'$ variable for a unit change in $'x'$ [11]. So, if instrument A measures $'y'$, and instrument B measures $'x'$ and if $y = x$, the slope of the straight line equation is equal to one. It is true that the straight line of $y = x$ will always have slope of 1. However, this is not always true in reverse, because for a line with a slope of 1, the straight line could be $y = x$, or could be $y = \alpha + x$. Therefore, testing the slope is equal to 1, is also an inappropriate method of testing agreement. When the test of slope is equal to 1 is significant, some people proceed to test the y-intercept. Theoretically, if slope is 1 and y-intercept is 0, then y will be equal to x ($y = x$). However, testing both slope and intercept to assess agreement is not so popular compared to other methods.

Bland and Altman [36] also suggested that the old measurement (y) can be regressed on the new measurement (x), and then one can calculate the standard error of a prediction of the old value from the new. This can be used to estimate predicted value from old measurements for any observed value of new measurement, with a confidence interval, which is also known as a *prediction interval* [36]. However, the problem is that the prediction interval is not constant; it is smaller in the middle, and wider towards the extremes [36].

4. Most commonly used methods to assess reliability

Various methods have also been used to estimate reliability. Recent review of reliability study in medicine found that among popular methods used include: intra-class correlation coefficient, comparing means, Bland-Altman limits of agreement, and correlation coefficient (r).

4.1. Intra-class correlation coefficient

The intra-class correlation coefficient (ICC) is the most popular method used to assess reliability of medical instruments [13]. The ICC was originally proposed by Fisher [37, 38]. He was a statistician from England, and Fisher's exact test was one of his well-known contributions to statistics [37, 39]. The earliest ICCs were modifications of the Pearson correlation coefficient [34]. However, the modern version of ICC is now calculated using variance estimates, obtained from the analysis of variance or ANOVA, through partitioning of the total variance between and within subject variance [14].

The general formula for ICC is given as [34]:

$$\text{ICC} = \frac{\text{Subject variability } (\delta_S^2)}{\text{Subject variability } (\delta_S^2) + \text{Mmeasurement error } (\delta_E^2)} \quad (7)$$

Values obtained from ANOVA table:

Measurement error, δ_E^2 = Mean square of Error, $M\,S_E$

$$\text{Subject variability, } \delta_S^2 = \frac{\text{Mean square of Ssubject, } M\,S_S - \text{Mmean square of Eerror, } M\,S_E}{\text{Number of observer}} \quad (8)$$

There is no ordering of the repeated measures and can be applied to more than two repeated measurements [5]. ICC is a ratio of variances derived from ANOVA, so it is unit less. The ratio is closer to 1.0, the higher the reliability [34]. Suppose, for example, that we measure carbon monoxide level for 10 patients using a same instrument three times. The hypothetical data are shown in **Table 6**. From the data, an ANOVA table can then be developed as in **Table 7**.

Patient	First reading	Second reading	Third reading	Mean
1	6	7	8	7
2	4	5	6	5
3	2	2	2	2
4	3	4	5	4
5	5	4	6	5
6	8	9	10	9
7	5	7	9	7
8	6	7	8	7
9	4	6	8	6
10	7	9	8	8

Table 6. Hypothetical data of repeated measurements of carbon monoxide level.

Source of variation	Sum of squares	Degree of freedom	Mean square
Patients	114	9	12.67
Raters/instrument	20	2	10
Error	10	18	0.56
Total	144	29	

Table 7. Analysis of variance summary table.

From **Table 7**, the value of ICC can be calculated:

$$\begin{aligned}\text{Measurement error, } \delta_E^2 &= MS_E = 0.56\\ \text{Subject variability, } \delta_S^2 &= \frac{MS_S - MS_E}{\text{Number of observer}}\\ &= \frac{12.67 - 0.56}{3} = 4.04\\ \text{ICC} &= \frac{4.04}{4.04 + 0.56} = 0.88\end{aligned} \tag{9}$$

The interpretation is that 88% of the variance in the measurements results from the 'true' variance among patients. However, note that this is according to the 'classical' definition of reliability. There are different forms of ICC depending on various assumptions or criteria. Chinn [40] recommended that any measure should have an intra-class correlation coefficient of at least 0.6 to be useful [40], whereas Rosner [41] suggested the interpretation of ICC as shown in **Table 8**.

ICC value	Interpretation
<0.4	Poor reliability
$0.4 \leq ICC < 0.75$	Fair to good reliability
≥0.75	Excellent reliability

Table 8. Interpretation of ICC.

4.2. Comparing means/mean difference

Second most popular method that has been used to assess reliability is to compare means of two sets of measurements (either using t-test or looking at the mean difference) [13]. Since reliability involves repeated measurement of the same subject, a paired t-test is usually applied. However, the paired t-test only gives information about differences between the means of two sets of data, and not about individual differences [14]. As in the explanation in Section 3.3, on assessing agreement, comparing means is also not a suitable method of assessing reliability.

4.3. Bland-Altman method

The Bland-Altman limits of agreement (LoA) also has been used as a method to assess reliability. Bland and Altman [20] suggested that LoA are suitable for the analysis of repeatability of a single measurement method. However, the use of LoA to evaluate reliability has been criticised, as it only estimates reliability when there are two observations for each subject [20]. This breaches the concept of reliability, that allows repeated (more than two) numbers of observations per subject [11]. Although Bland and Altman [42] suggested methods to deal with multiple measurements in calculating the LoA, this method is more suitable for the analysis of agreement rather than reliability. They proposed calculating the mean of the replicated measurements by each instrument, for each subject [42]. Then, these pairs of means could be used to compare the two instruments using the limits of agreement [42]. The use of LoA in the analysis of reliability also has been criticised by Hopkins [43], who gave reasons why LoA is not the best method to use for reliability analysis. According to Hopkins [43], the values of the LoA can result in up to a 21% bias, and this depends on the degrees of freedom of the reliability study (i.e. number of participants and trials).

4.4. Correlation coefficient

As discussed in Section 3.2, the correlation coefficient provides information about the association and the strength of linear relationship. Correlation will not detect any systematic or fixed errors, and it is possible to have two sets of scores that are highly correlated, but not repeatable [14]. Therefore, it is recommended that the correlation coefficient should not be used in isolation for measuring reliability [14, 44]. Furthermore, the correlation coefficient also breaches the concept of reliability, as it only estimates reliability when there are only two observations for each subject [11].

5. Summary

Agreement signifies the accuracy of certain instruments, whereas reliability indicates precision. It is imperative that all medical instruments are accurate and precise. Otherwise, a failure may lead to critical medical errors. Therefore, there is a necessity for the proper evaluation of all medical instruments, and it is important to be sure that the appropriate statistical method has been used. Preferably, agreement and reliability should be assessed together in a validation study.

Simplicity, practicality (or interpretability) and ability of a certain method to detect systematic bias are among the important factors when choosing a method to evaluate agreement. Nonetheless, the ability of the method in detecting bias is still the priority. The use of multiple methods has the advantage of compensating for the limitations of any single method. However, the application of multiple inappropriate statistical methods should be avoided.

Finally, the inappropriate analysis in the method comparison study is a cause for concern in the medical field and cannot be ignored. It is important for medical researchers and clinicians from all specialties to be aware of this issue because inappropriate statistical analyses will lead to inappropriate conclusions. Thus, jeopardising the quality of the evidence, which may in turn, influences the quality of care given to the patients. Educating medical researchers on methods in validation study and clear recommendations and guidelines on how to perform the analysis will improve their knowledge in this area, and help reducing the problem of inappropriate statistical analysis. It is also important to involve statisticians who are able to understand in depth of various statistical methods in the medical education program. Consulting statisticians or inviting them as part of the medical research team also could help in reducing mistake in statistical analysis.

Acknowledgements

I would to express my utmost gratitude to Prof. Dr. Awang Bulgiba and Prof. Dr. Noor Azina Ismail from the University of Malaya for their support, guidance and assistance when I was conducting research in this area. Also for their motivation and support in my career development.

Author details

Rafdzah Zaki

Address all correspondence to: rafdzah@hotmail.com

Department of Social & Preventive Medicine, Faculty of Medicine, Julius Centre University of Malaya, Kuala Lumpur, Malaysia

References

[1] Portney, L.G, Watkins, M.P. Foundations of clinical research: Applications to practice. New Jersey: Prentice-Hall; 2000.

[2] Feinstein, A.R. Clinimetrics. New Haven: Yale University Press; 1987.

[3] de Vet, H.C.W, Terwee, C.B, Bouter, L.M. Current challenges in clinimetrics. Journal of Clinical Epidemiology. 2003;**56**(12):1137–1141.

[4] de Vet, H.C.W, Terwee, C.B, Bouter, L.M. Clinimetrics versus psychometrics: two sides of the same coin. Journal of Clinical Epidemiology. 2003;**56:**1146–1147.

[5] Streiner, D.L, Norman, G.R. Health measurement scales. Apractical guide to their development and use. 2nd ed. Oxford: Oxford University Press; 1995.

[6] Haynes, S.N, Richard, D.C.S, Kubany, E.S. Content validity in psychological assessment: a functional approach to concepts and methods. Psychological Assessment. 1995;7:238–247.

[7] Baxter, P. Invalid measurement validity. Developmental Medicine & Child Neurology. 2005;**47**:291.

[8] Lambert, H.C, Gisel, E.G, Wood-Dauphinee, S. The functional assessment of dysphagia: psychometric standards. Physical & Occupational Therapy in Geriatrics. 2002;**19**:1–14.

[9] Innes, E, Straker, L. Validity of work-related assessments. Work. 1999;**13**:125–152.

[10] de Vet, H.C.W, Terwee, C.B, Knol, D.L, Bouter, L.M. When to use agreement versus reliability measures. Journal of Clinical Epidemiology. 2006;**56**:1033–1039.

[11] Fay, M.P. Random marginal agreement coefficients: rethinking the adjustment for chance when measuring agreement. Biostatistics. 2005;**6**(1):171–180.

[12] Zaki, R, Bulgiba A, Ismail R, Ismail N.A. Statistical methods used to test for agreement of medical instruments measuring continuous variables in method comparison studies: a systematic review. PLoS One. 2012;**7**(5):e37908. DOI: 10.1371/journal.pone.0037908

[13] Zaki, R, Bulgiba, A, Nordin, N, Ismail, N.A. A systematic review of statistical methods used to test for reliability of medical instruments measuring continuous variables. Iranian Journal of Basic Medical Science. 2013;**16**(6):803–807.

[14] Bruton, A., Conway, J.H, Holgate, S.T. Reliability: what is it, and how is it measured? Physiotherapy. 2000;**86**(2):94–99.

[15] Altman, D.G, Bland, J.M. Measurement in medicine: the analysis of method comparison studies. The Statistician. 1983;**32**:307–317.

[16] Bahareh, A, Saeed, H, Ramin, H. Comparison of Harris benedict and Mifflin-St Jeor equations with indirect calorimetry in evaluating resting energy expenditure. Indian Journal of Medical Science. 2008;62(7): 283–290.

[17] Daly, L.E, Bourke, G.J. Interpretation and use of medical statistics. 5th ed. Oxford: Blackwell Science Ltd.; 2000.

[18] Ahn, Y, Garruto, R.M. Estimations of body surface area in newborns. Acta Paediatrica (Oslo, Norway: 1992). 2008;**97**(3):366–370.

[19] Milias, G.A, Antonopoulou, S, Anthanasopoulos, S. Development, reliability and validity of a new motorized isometric dynamometer for measuring strength characteristics of elbow flexor muscles. Journal of Medical Engineering & Technology. 2008;**32**(1):66–72.

[20] Bland, J.M, Altman, D.G. Statistical methods for assessing agreement between two methods of clinical measurement. Lancet. 1986;1(8476):307–310.

[21] Bland, J.M, Altman, D.G. Agreed statistics: measurement method comparison. Anesthesiology. 2012;**116**(1):182–185.

[22] Hopkins, W.G. Bias in Bland-Altman but not regression validity analyses. Sportscience. 2004;**8**:42–46.

[23] Zaki, R, Bulgiba, A, Ismail, N.A. Testing the agreement of medical instruments: overestimation of bias in the Bland-Altman analysis. Preventive Medicine. 2013;**57**(Suppl):S80–S82.

[24] Nazir, Z, et al. Revisiting the accuracy of peak flow meters: a double-blind study using formal methods of agreement. Respiratory Medicine. 2005;**99**:592–595.

[25] Luiz, R.R, Szklo, M. More than one statistical strategy to assess agreement of quantitative measurements may usefully be reported. Journal of Clinical Epidemiology. 2005;**58**(4):215–216.

[26] Lee, J, Koh, D, Ong, C.N.Statistical evaluation of agreement between two methods for measuring a quantitative variable. Computers in Biology and Medicine. 1989;**19**(1):61–70.

[27] Lin, L.I. Total deviation index for measuring individual agreement with applications in laboratory performance and bioequivalence. Statistics in Medicine. 2000;**19**(2):255–270.

[28] Bland, J.M. An Introduction to medical statistics. 2nd ed. Oxford: Oxford University Press; 1995.

[29] Wikipedia. The Free Encyclopedia. Correlation [Internet]. Available from: http://en.wikipedia.org/wiki/Correlation. [Accessed: 30 January 2009]

[30] Hanbazaza, S.M, Mansoor, I. Accuracy evaluation of point-of-care glucose analyzers in the Saudi market. Saudi Medical Journal. 2012;**33**(1):91–92.

[31] Mårtensson, M, Winter, R, Cederlund, K, Ripsweden, J, Mir-Akbari, H, Nowak, J, Brodin, L. Assessment of left ventricular volumes using simplified 3-D echocardiography and computed tomography – a phantom and clinical study. Cardiovascular Ultrasound. 2008;**6**(26)DOI: 10.1186/1476-7120-6-26.

[32] Bland, J.M, Altman, D.G. A note on the use of the intraclass correlation coefficient in the evaluation of agreement between two methods of measurement. Computers in Biology and Medicine. 1990;**20**(5):337–340.

[33] Shrout, P, Fleiss, J. Intraclass correlations: uses in assessing rater reliability. Psychological Bulletin. 1979;**86**(2):420–428.

[34] Weir, J.P. Quantifying test-retest reliability using the intraclass correlation coefficient and the SEM. Journal of Strength and Conditioning Research. 2005;**19**(1):231–240.

[35] McGraw, K, Wong, S. Forming inferences about some intraclass correlation coefficients. Psychological Methods. 1996;**1**:30–46.

[36] Bland, J.M, Altman, D.G. Applying the right statistics: analyses of measurement studies. Ultrasound in Obstetrics and Gynecology. 2003;**22**(1):85–93.

[37] Wikipedia. The Free Encyclopedia. Ronald A Fisher [Internet]. Available from: https://en.wikipedia.org/wiki/Ronald_A._Fisher. [Accessed: 12 June 2011]

[38] Fisher, R.A. Statistical methods for research workers. Edinburgh: Oliver and Boyd; 1925.

[39] Fisher, J. R. A. Fisher: The life of a scientist. New York: Wiley; 1978.

[40] Chinn, S. Repeatability and method comparison. Thorax. 1991;**46**:454–456.

[41] Rosner, B. Fundamentals of biostatistics. 6th ed. Duxbury: Thomson Brooks/Cole; 2006.

[42] Bland, J.M, Altman, D.G. Measuring agreement in method comparison studies. Statistical Methods in Medical Research. 1999;**8**(2):135–160.

[43] Hopkins, W.G. Measures of reliability in sports medicine and science. Sports Medicine. 2000;**30**(1):1–15.

[44] Neveu, D, Aubas, P, Seguret, F, Kramar, A, Dujols, P. Measuring agreement for ordered ratings in 3 × 3 tables. Methods of Information in Medicine. 2006;**45**(5):541–547.

11

Signal Optimal Smoothing by Means of Spectral Analysis

Guido Travaglini

Abstract

This chapter introduces two new empirical methods for obtaining optimal smoothing of noise-ridden stationary and nonstationary, linear and nonlinear signals. Both methods utilize an application of the spectral representation theorem (SRT) for signal decomposition that exploits the dynamic properties of optimal control. The methods, named as SRT1 and SRT2, produce a low-resolution and a high-resolution filter, which may be utilized for optimal long- and short-run tracking as well as forecasting devices. Monte Carlo simulation applied to three broad classes of signals enables comparing the dual SRT methods with a similarly optimized version of the well-known and reputed empirical Hilbert-Huang transform (HHT). The results point to a more satisfactory performance of the SRT methods and especially the second, in terms of low and high resolution as compared to the HHT for any of the three signal classes, in many cases also for nonlinear and stationary/nonstationary signals. Finally, all of the three methods undergo statistical experimenting on eight select real-time data sets, which include climatic, seismological, economic and solar time series.

Keywords: signal analysis, white and colored noise, Monte Carlo simulation, time series properties, smoothing techniques

1. Introduction

The literature on time series smoothing and denoising techniques is vast and encompasses different disciplines, such as chemometrics, econometrics, seismology, signal analysis and many more. Among the most renown and utilized methods are the Savitzky-Golay and Hodrick-Prescott filters, the Hilbert-Huang transform (HHT), wavelet analysis, as well as the ample class of kernel filters [1–6].

All of these techniques share an ample spectrum of degrees of resolution loss of the original raw signal, namely, a large family of filters addressed at separating from noise the stochastic or broken trend underlying the observed signal. The researcher is thus enabled to select the desired denoising frequency visually and/or by means of some prior information, like threshold selection, rescaling and data-dependent smoothing weights. Automatic selection procedures are available in the context of Savitzky-Golay filtering developed in the field of chemometrics [7] and quite a few in the field of image processing (e.g. [8]). Such procedures are utilized also for the stoppage criteria embedded in the HHT but are virtually absent in the field of econometrics [3, 8–11].

In general, however, the capability of balancing high- with low-resolution characteristics in a world of different real-time data (RTD) patterns is oftentimes questionable. In fact, in spite of their efficacy, all of these models contribute a common shortcoming by letting the researcher arbitrarily—and thus casually—select the degree of resolution, thereby risking over/under fitting of the slow mode with respect to the original signal. The immediate consequence is suboptimal denoising, namely, a signal extraction yielding a smoother that is statistically inconsistent with the original peak/trough pattern or that too closely replicates the original signal. For this purpose, automated and optimizing parameter selection must be utilized, based on the minimum squared distance between the actual and the smoothed series [12].

The goal of this chapter exactly addresses this problem, which may be defined as the search for statistical dynamic efficiency of the estimated smoother. This search, in practice, ends up with extracting from the available data set the optimal smoother, namely, the stochastic or broken trend, which minimizes the noise variance among many second-best solutions. Section 2 introduces time series decomposition and the statistical taxonomy of stochastic noise. Sections 3 and 4 are devoted to an introduction and detailed description, respectively, of the HHT and of the SRT models for signal smoothing. Section 5 produces comparative efficiency results of the two new techniques with respect to HHT by using Monte Carlo simulations and reports descriptive statistics of some select climatic, seismological, economic and solar real-time series. Section 6 concludes, while the Appendix produces the sources of the real-time signals utilized.

2. Signal decomposition and statistical taxonomy, linearity and stationarity testing

Any observed signal, by means of additive decomposition [3, 13], may be modeled as the sum of three different random variables of unknown distribution such that

$$Y_t = y_t^* + s_t + y_t \tag{1}$$

where Y_t is the time sequence of the real-valued observations for the discrete-time notation $t \in [1, T]$, $T < \infty$. Moreover, $y_t^* = \mathrm{E}\left(Y_t \middle| \Omega_{t-j}\right)$ where $\Omega_{t-j} = \left\{Y_{t-j}\right\}_j^J$ is the information set available at $t-j$, $j \in [1, J]$ for $J \leq T$ a maximum preselect lag [14, 15]. The first term of Eq. (1)

is the slow-mode or growth component in the form of a broken trend [16] or of a smoother [3]. The second component s_t is the periodical seasonal cycle, if any and the last component is the fast-mode component where $\mathrm{E}(y_t) \sim \mathrm{IID}(0, \sigma_y^2)$.

The signal may be linear or nonlinear, as well as stationary or nonstationary, depending on its components and on their distributional properties. Briefly, nonlinearity is a feature of a signal characterized by the presence of linear as well as quadratic and/or cubic variables and/or multiplicative variables, rendering the signal typically nonparametric and thus unsuitable for standard statistical testing such as variance analysis, prediction and stationarity. In addition, a linear or nonlinear signal is stationary (nonstationary) if its best linear trend fit is significantly flat (rising/falling over time). Nonlinearity and nonstationarity of the signal [17] can be tested by means of appropriate procedures [18–21]. The resulting number of regime switches, when applicable, is determined by computing single or multiple intercept and/or trend time breaks [22–24]. For details on the corresponding critical-value statistics, the reader is redirected to the respective authors.

All components of the observed signal are stochastic and unobservable and must be retrieved from the raw data set by means of appropriate computational methods, some of which were invoked in Introduction. In particular, the slow-mode term requires the use of some filtering procedures to possibly attain an optimal low-resolution time-varying trend, in fact the smoother (Section 3). In addition, if s_t is a high- or low-resolution phenomenon, it may be entrenched into either or y_t^* and thus difficult to disentangle from either component.

Any observed signal in Eq. (1) is shown to pertain to one of the following three statistical taxonomic classes: (1) a random Gaussian white noise (RGS), (2) a colored (pink/red) random noise (RWS) and (3) a valid-threshold regime-switching Gaussian mixture constructed as a casually ordered sequence of two or more of the previous signals (RGWS).

Formally, the first kind of signal (RGS) may be expressed as follows:

$$Y_t = c + \varepsilon_t + \delta t \tag{2}$$

where the constant term, which corresponds to the mean of the noise and the time-slope coefficient, respectively, are $c,\ \delta \in (-\infty, \infty)$ and $\{\varepsilon_t\}_{t=1}^{T}$ is a real-valued univariate white-noise data sequence such that $\varepsilon_t \sim \mathrm{IID}(0, \sigma_\varepsilon^2)$ and $\sigma_\varepsilon^2 << \infty$ is a constant. In Eq. (2), if $\delta = 0$, the noise is trend-stationary and the process is a pure RGS with constant mean and variance and expected zero-autocovariance. Instead, if significantly $\delta \neq 0$, the noise is trend-nonstationary and exhibits a rising/falling trend line with white noise superimposed. This result may be substituted into Eq. (1) such that the slow model, and with it the entire signal, would be significantly trended.

The second kind of signal (RWS) is

$$Y_t = c + \sum_{t=1}^{T} \varepsilon_t + \delta t \tag{3}$$

where the coefficients contribute the same characteristics as those in Eq. (2), but even if $\delta = 0$ the noise is trend-nonstationary and both the variance and the autocovariance are not constant overtime. In consequence, the process, which is an additive white noise, is entirely time dependent in both mean and variance and this affects the overtime pattern of the signal and its components.

The third kind of signal (RGWS) is represented by the Markov regime-switching model [25–31], where the raw signal may undergo significant structural changes across its lifetime due to variability in its underlying probability transition matrix. Therefore, the signal, which is a combination of Eqs. (2) and (3), may experience some quiet and some turbulent states of nature, like stock-market fluctuations and seismic signals. The obvious consequence is that the signal may fall short of standard parametric normality or stationarity statistical testing. Moreover, a threshold upper and lower limit must be imposed on regime-switching dates to avoid spurious computation of clusters too close to each endpoint. The threshold is set in the present context to be 15–85% of the total number of observations with consequential "valid" dates comprised within the reduced-size sample [24].

The RGWS noise class is represented by the following sequence

$$Y_{t,m} = \begin{Bmatrix} c_1 + f_1\left(\varepsilon_{t(1),1}\right) + \delta_1 t(1) \\ \vdots \\ c_m + f_m\left(\varepsilon_{t(m),m}\right) + \delta_m t(m) \end{Bmatrix} \tag{4}$$

where $m \geq 2$ are the valid state indexes and, for $i \in [1, m]$, the time index $t(i)$ is state-specific, the expected value $E\left(\left|\delta_i\right|\right) \geq 0$, whereas $f_i(.)$ are the nonadditive or additive white-noise functions, respectively, from Eqs. (2) and (3). From Eq. (4), the typical ith state of the signal is

$$Y_{t,i} = c_i + f_i\left(\varepsilon_{t(i),i}\right) + \delta_i t_i;\ \forall i \in m \tag{5}$$

where, given an m-sized vector of exponents represented by a sequence of positive integers $1 \leq \alpha(i) \ll \infty$, the fast modes of the signal are

$$\begin{cases} f_i\left(\varepsilon_{t(i),i}\right) = \left(\varepsilon_{t(i)}\right)^{\alpha(i)}, \text{ for some } 1 \leq i < m \\ f_i\left(\varepsilon_{t(i),i}\right) = \left(\sum_{t(i)=1}^{T(i)} \varepsilon_{t(i)}\right)^{\alpha(i)}, \text{ for the remaining } 1 \leq i < m \end{cases} \tag{6}$$

The entire m-sized sequence of the state functions $f_i\left(\varepsilon_{t(i),i}\right)$ may turn to be highly nonlinear if at least one $\alpha(i) > 1$. As a result, the entire signal would give rise in many cases to quirky graphics characterized by enhanced peaks and/or troughs, possibly intermitted by sizably flatter states. Multiple regime switching is expected to be very frequent in such a case, much less so if $\alpha(i) \equiv 1$, where Eq. (4) would show a more moderate pattern and fewer state gyrations, even zero.

Either of the three taxonomies enables empirical estimation of Eq. (1) by means of any of the smoothing procedures introduced in Section 1. Such procedures in different manners will find the smoother y_t^* and by default the component y_t. In the IID context applied to the noise ε_t of Eq. (2) to Eq. (4), the random slow and fast components of Eq. (1) are both discrete stochastic processes with zero or nonzero mean and with finite variance and autocovariance. Incidentally, if the slow component is significantly trended (not trended), its autocovariance process is persistent (tapers off slowly over time), whereas the autocovariance of the second component rapidly tapers off over time (is white noise).

Whereas by simple logic the extraction process of the smoother y_t^* is carried out over the entire timespan of the signal, in the presence of significant regime changes in Eqs. (4) and (5), this procedure might not produce an optimal smoother. Therefore, extraction of the smoother may be applied on a subsample cluster basis rather than by full-sample computation. Each subsample corresponds to a state (Eq. (4)) and its own smoother is estimated separately from the others. The sequence so obtained is then tested by minimum percentage root mean squared error (PRMSE). In quite, a few real-time data cases (Section 5), the subsample estimation technique is found to be a better performer.

As a matter of empirical fact, simulation outcomes of Monte Carlo iterations with 1000 draws indicate that around 60% of the RGWS signals with $\alpha(i) \equiv 1$ in Eq. (5) are nonlinear, while on average, roughly 40% are stationary and almost half of them contain one or more valid regime switches. Nonlinearity and stationarity are very common in the RGWS signals with $\alpha(i) > 1$ in Eq. (5), as they are found in 70% of the corresponding simulations. Finally, on average, less than 1% (10%) of the RGS (RWS) are nonstationary (stationary), whereas in most cases (over 90%), both signals are linear.

3. The Hilbert-Huang transform (HHT)

The Hilbert-Huang transform (HHT) is an empirical procedure designed to correct for noise-ridden signal smoothing, purported to work for both nonlinear and nonstationary time series [4–6, 32, 33]. The HHT is based on the so-called ensemble empirical mode decomposition (EEMD), which consists, for any arbitrary number of steps $2 \leq Q \leq T$, of producing at each step an intrinsic mode function (IMF). Each IMF is computed by consecutive "siftings" of the data, a procedure that involves finding the means of the high-resolution envelopes constructed by cubic splining of the extreme values of the available data, possibly after correcting for endpoints [34]. The Q-step IMF-sifting process for a given signal Y_t is represented by the following one-lag adaptive sequence

$$\begin{aligned} h_{t,0} &= Y_t \\ h_{t,1} &= h_{t,0} - \mathrm{E}\left(h_{t,0}\right) \\ h_{t,2} &= h_{t,1} - \mathrm{E}\left(h_{t,1}\right) \\ &\dots \\ h_{t,Q} &= h_{t,Q-1} - \mathrm{E}\left(h_{t,Q-1}\right) \end{aligned} \tag{7}$$

where, given the upper and lower envelopes $s^U_{t,p}$, $s^L_{t,p}$, $p \in [0, Q-1]$, $\mathrm{E}(h_{t,p})$ is the mean envelope obtained at the end of each sifting process. More generally, from the second line onward, Eq. (7) may be written as follows:

$$h_{t,q} = h_{t,q-1} - \mathrm{E}(h_{t,q-1}),\ q \in [1, Q] \tag{8}$$

which represents the family of all the sifted IMFs starting from the highest frequency. Subsequently, the matrix $HS_{T,Q}$:(T, Q) of the HHT smoothers is obtained. This is the matrix of the EEMDs. Its rows are expressed as $\tilde{Y}_{t,q} = h_{t-q} - h_{t-q-1}$ from which from which the noise estimates are

$$u_{t,q} = \tilde{Y}_{t,q} - \tilde{Y}_{t,q-1},\ u_{t,q} \sim \mathrm{IID}(0, \sigma^2_u). \tag{9}$$

In an HHT environment, the optimal high-resolution smoother among the candidates of $HS_{T,Q}$ may be detected by utilizing the stoppage criterion for sifting, or else by similar means closely akin to the percentage root mean squared error (PRMSE), a much-used performance index for goodness-of-fit purposes. The PRMSEs may be computed as follows:

$$S_q = \left(T^{-1} \frac{\sum_{t=1}^{T} u^2_{t,q}}{\sum_{t=1}^{T} \tilde{y}^2{}_{t,q-1}} \right)^{0.5},\ \forall q \in Q \tag{10}$$

which produce an inverse signal-to-noise ratio screeplot of length Q [35, 36]. The procedure for detecting the screeplot global minimizer, denoted as Q^*, requires

$$P_q = \frac{\|S_{q-2} - S_{q-1}\|}{\|S_{q-1} - S_q\|} \tag{11}$$

where $\|.\|$ is the Euclidean norm of the enclosed argument and $Q^* = \arg\max_{1 \le q \le Q}(P_q)$ [37]. There immediately follows that the measured PRMSE, given Q^*, is represented by the following formula

$$\hat{S}_{Q^*} = \left(T^{-1} \frac{\sum_{t=1}^{T} u^2_{t,Q^*}}{\sum_{t=1}^{T} \tilde{y}^2_{t,Q^*}} \right)^{0.5} \tag{12}$$

which is the flagship of the efficiency indicators that shall be utilized on confrontational grounds with respect to the two SRTs discussed in Section 4.

4. The spectral representation transform (SRT)

By virtue of the spectral representation theorem, the signal Y_t as in Eq. (1) may be approximated by the following De Moivre's formula, as follows:

$$Y_t \simeq \cos(\omega t) - r * \sin(\omega t) \tag{13}$$

which depicts a harmonic waveform continuous function, where $r = \sqrt{-1}$ and ω are a given arbitrary and constant frequency. There follows that the signal Y_t may be defined by the following periodic function

$$\hat{Y}_{t,k} \equiv \mu + \sum_{k=1}^{K} \left[\phi_k \sin\left(\omega_k \cdot (t-1)\right) + \varphi_k \cos\left(\omega_k \cdot (t-1)\right) \right] \tag{14}$$

where μ is the mean of the signal, $k \in [2 \leq K \leq T)$ for K a maximum lag integer, $\{\phi_k\}_{k=1}^{K}$, $\{\varphi_k\}_{k=1}^{K}$ are real-valued random coefficient sequences, both IID with finite mean and variance and $\{\omega_k\}_{k=1}^{K}$ is the frequency sequence such that

$$\omega_k = \begin{cases} T^{-1} 2\pi k \text{ if } \Delta Y_t = f(Y_{t-1}), \ \lim\limits_{k \to 0.5T}(\omega_k = \pi) \\ T^{-1} \pi k \text{ if } \Delta Y_t = f(Y_{t-1}^3), \ \lim\limits_{k \to T}(\omega_k = \pi) \end{cases} \tag{15}$$

where $\Delta y_t = f(.)$ expresses the existence of linearity and nonlinearity of the process, depending on the exponent attached to the lagged signal level [20, 21]. Needless to say, K generally corresponds to the maximal EMD level contemplated in the HHT method.

The fitted signal of Eq. (14), which is the smoother estimable by ordinary least squares [25], is actually the SRT of the original signal with time-varying amplitude and lag k. If we let the time series of the prediction error be

$$e_{t,k} = Y_t - \hat{Y}_{t,k}, \ \forall k \in K \tag{16}$$

where $e_{t,k} \sim \text{IID}\left(0, \sigma_{e,k}^2\right)$. After producing 1000 Monte Carlo normal simulations of Eq. (14) with $T = 300$, the central limit theorem (CLT), as $k \to T$, is found not to hold asymptotically for RGWS, as is obvious in the presence of regime switches. Elsewise, for RGS and RWS, we have $\lim\limits_{k \to T}\left(Y_t = \hat{Y}_{t,k}\right)$.

Similar to the technique utilized for identifying the optimal smoother in the PRMSE sense exhibited in Section 3, also here a matrix of smoother candidates may be obtained depending on the kth lag chosen. The matrix is defined as $S\,S_{T,K}\!:(T, K)$ from which the optimal smoother can be selected by applying the optimal lag stopping criterion similar in kind to Eq. (16). However, we utilize here a performance index for model selection different from the PRMSE, which is based on the dynamics of the Hamiltonian optimal control problem [38–40].

If we let $e_{t,k}$ be defined as in Eq. (16), then its dynamics is captured by the first-order autoregressive AR(1) process $g_{t,k} = e_{t,k} - e_{t-1,k}$ while the dynamics of the smoother is expressed as the AR(1) process $h_{t,k} = \hat{Y}_{t,k} - \hat{Y}_{t-1,k}$. We expect both processes to be normally distributed with zero mean and finite variance.

Hence, the Hamiltonian problem may be expressed as follows:

$$H_k = \frac{1}{2} \sum_{t=2}^{T} \left\{ e_{t,k}^2 + g_{t,k}^2 + h_{t,k}^2 \right\}, \ \forall k \in K \tag*{17)}$$

where the first element within the curly braces is of obvious reading and represents a resolution index that picks up the high frequencies of the problem. The second and the third elements capture the AR(1) dynamics of the processes involved in the Hamiltonian problem, namely, those of the prediction error and those of the signal itself. Eq. (17) is a cubic spline smoother that partly resembles the Hodrick-Prescott filter in the inclusion of both a cyclical and a trend part [3]. Moreover, Eq. (17) forms the basis for a screeplot [36] similar in kind to that of Eq. (11), whereby the optimal lag K^* is obtained after letting

$$V_k = \frac{\|H_{k-2} - H_{k-1}\|}{\|H_{k-1} - H_k\|} \tag{18}$$

such that $K^* = \arg\max_{1 \le k \le K}(V_k)$ and wherefrom $\hat{Y}^*_{t,K^*}$ is the optimal low-resolution Hamiltonian-based smoother, which is named SRT1.

From this smoother the dual high-resolution SRT smoother, named SRT2, may be easily constructed by means of envelope augmentation, a procedure knowingly embedded into the HHT sifting process (Section 3). Let the upper and the lower envelopes of the actual signal be defined as the unique cubic splines s_t^U, s_t^L constructed by using the extrema of the given data. The upshot is the signal $\hat{Y}^{**}_{t,K^*} = \frac{1}{2}(s_t^U + s_t^L) \in [s_t^U,\ s_t^L]$, such that the second smoother is $\tilde{Y}^*_{t,K^*} = \frac{1}{2}\left(\hat{Y}^*_{t,K^*} + \hat{Y}^{**}_{t,K^*}\right)$, which is the mean of the two smoothers. The error time series, for the optimal given lag K^*, are expressed in a fashion similar to Eq. (9), as follows

$$\begin{cases} \varepsilon_{t,K^*} = Y_t - \hat{Y}^*_{t,K^*}, & \varepsilon_{t,K^*} : \text{IID}(0, \sigma_\varepsilon^2) \\ \eta_{t,K^*} = Y_t - \tilde{Y}^*_{t,K^*}, & \eta_{t,K^*} : \text{IID}(0, \sigma_\eta^2) \end{cases} \tag{19}$$

where the first error is associated with SRT1 and the second is associated with SRT2.

From Eq. (19), the PRMSEs of both models may be found by the same means as those employed to obtain Eq. (12), namely

$$\hat{W}_{K^*} = \left(T^{-1}\frac{\sum_{t=1}^{T}\varepsilon^2_{t,K^*}}{\sum_{t=1}^{T}\hat{Y}^2_{t,K^*}}\right)^{0.5}, \quad \tilde{W}_{K^*} = \left(T^{-1}\frac{\sum_{t=1}^{T}\eta^2_{t,K^*}}{\sum_{t=1}^{T}\tilde{Y}^2_{t,K^*}}\right)^{0.5} \tag{20}$$

where the first (second) index is associated with SRT1 (SRT2).

5. Smoother analysis applied to artificial and real-time signals

Figure 1 depicts six signals pertaining to the three taxonomic classes (Section 2). The first three signals are 300 observation artificial RGS, RWS and RGWS, drawn from standard normal distributions. The other three are real-world signals, of which the first two are recent Japanese earthquake seismographic waveforms collected from a specific web directory [41]. The last signal represents the Qualcomm NASDAQ-traded stock close price on a weekly basis from

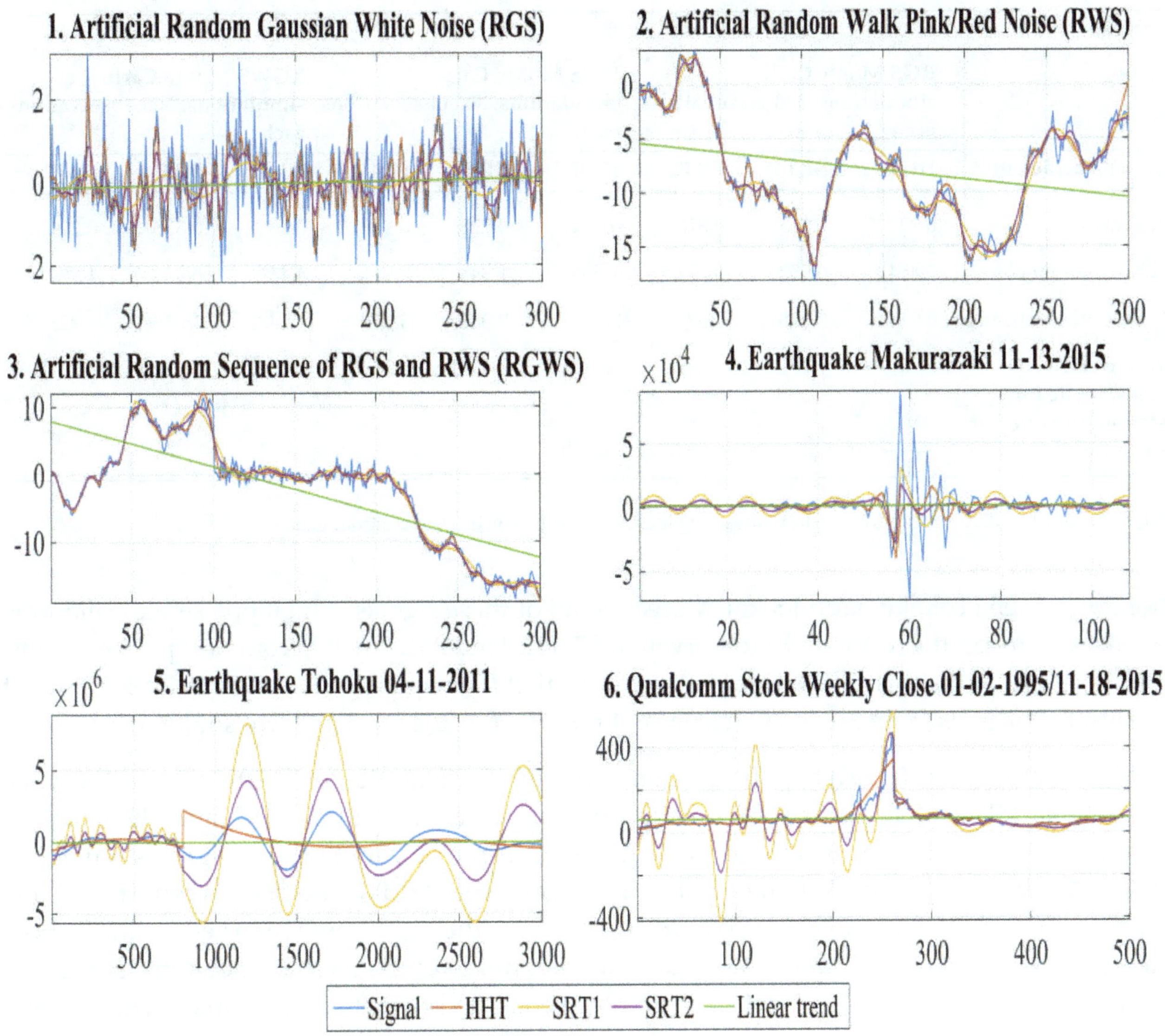

Figure 1. Artificial and real-time random signals, all smoothers and linear trends. Vertical axes represent signal magnitudes, horizontal axes represent time in calendar years (1, 2, 3 and 6) or in seconds (4, 5).

the year 1995 to date. In particular, with regard to the first two real-time signals (**Figure 1**, panels 4 and 5), the reported seismographs concern two earthquakes, respectively, occurred in Makurazaki (Kagoshima prefecture) on November 13, 2015 at 20:51 UTC, with moment magnitude (M_w) of 6.7 and no damages no victims and in the Tohoku region on March 07, 2011 at 5:46 UTC, with M_w of 9.0, an ensuing tsunami and a reported death toll of 15,893. The waveform data from respective seismogram stations include 108 and 47,200 observations of which only the last 3000 were retained for experimentation.

Horizontal (vertical) measurements in **Figure 1** represent time lengths (magnitudes). All of the three smoothers (HHT, SRT1 and SRT2) are included together with a linear best-fit trend. The first two signals (RGS and RWS) are linear while the third (RGWS) is nonlinear (**Figure 1**, panels 1–3). Among them, only RGS is stationary and none exhibits a valid time break. Of the three real-time signals (Makurazaki and Tohoku earthquakes, Qualcomm), all are nonlinear and one is nonsta-

Tables	A. RGS Monte Carlo simulations, 300 observations each			B. RWS Monte Carlo simulations, 300 observations each			C. RGWS Monte Carlo simulations, 300 observations each		
Signal/coefficient	HHT	SRT1	SRT2	HHT	SRT1	SRT2	HHT	SRT1	SRT2
1. PRMSE	0.924	0.973	0.875	0.224	0.228	0.164	0.288	0.355	0.221
2. Error variance	0.854	0.952	0.771	1.633	1.844	0.911	4.459	5.533	1.973
3. Smoother variance	0.924	0.978	0.881	7.290	8.088	5.554	15.482	15.586	8.928
4. Corrected ADF test statistic for stationarity of error variance	-3.138	1.472	-6.851	-4.088	-3.243	-4.067	-1.831	-1.565	-3.104

Table 1. Monte Carlo simulation mean values of select coefficients for the three signal classes.

tionary as well (Tohoku earthquake). Moreover, all of these signals exhibit one single valid time break each, respectively, located at observation 57, 1123 and 261. Finally, from unreported results, SRT2 prevails as the least-PRMSE smoother for the first four signals, while the quirky behavior of the last two signals (Tohoku and Qualcomm) finds HHT as the most efficient smoother.

Table 1, columns A–C, exhibits the efficiency performance and other coefficients of the three smoothers (HHT, SRT1 and SRT2) for the three signal classes (RGS, RWS and RGWS). The results shown are the mean values obtained from empirical Monte Carlo signal simulations with 1000 draws each and length of 300 observations. The PRMSE coefficients reported in line 1 have received attention in Eqs. (12) and (20), whereas the coefficients reported in line 2 show the simulated mean error variance appearing in their numerators. The variance of the smoothed trends is exhibited in line 3. In all cases, simple eyeballing points to SRT2 as the most efficient, that is, the variance minimizing smoothing method (**Table 1**, third column of columns A–C).

Stationarity of the error variances in Eqs. (9) and (19) is crucial for establishing the goodness-of-fit of the estimated smoothers of all signal classes. Here, stationarity is tested by means of a novel technique, which corrects the conventional augmented Dickey-Fuller test statistic [19] after accounting for overtime changes in subsample variances, cycles and growing/falling linear trends in the signal [12]. This technique exhibits a similar nonparametric distribution as the ADF test statistic, such that the critical test statistics for stationarity of the untransformed signal are -2.79 and -2.51 for p-values of 1 and 5%, respectively. The corrected ADF test statistics of the error variances are reported in **Table 1**, line 4. In many, if not in most cases, stationarity emerges, yet the error variances generated by SRT2 are significantly stationary for all signal classes, whereas the other methods fail in the particular case of RGWS.

In order to further compare the HHT and the SRT performances, select real-time signals are being put to empirical testing for the sake of optimal smoothing analysis. The entire data set of the real-time data (RTD) includes eight signals of different nature: climatic, economic, seismic and solar. The RTD proposed are mostly high resolution and long term, ranging from a minimum of 315 to a maximum of 1915 observations of which one is yearly and all the others are monthly. The last access to the web for all available data was November 30, 2015.

The full source list and related time intervals are contained in the RTD Appendix, while the synthetic acronymed list of the select signals is the following: AMO (Atlantic Multidecadal Oscillation), GISS (Global Land-Ocean Temperature Index), Yamalia (Yamal Peninsula Temperature Reconstructions), S&P500 (Standard & Poor's 500 Composite Index), SPI (U.S. Standardized Precipitation Index), Banda (Banda Aceh earthquake, 2004), NASDAQ (Close values of the NASDAQ Stock Index) and finally SSN (Sunspot Numbers).

Among these RTDs, worth of more details than those provided in the Appendix is the earthquake of Banda Aceh, Indonesia, which occurred on December 26, 2014 at 00:58 UTC with M_w of 9.2 and which caused, especially due to an extraordinarily violent tsunami wave, a death toll of an estimated 250,000. The data utilized straddle the foreshock and the main shock tremors, namely, the observations comprised between 12,000 and 15,000 out of a total of recorded waves tallying 58,320 observations.

The relevant descriptive statistics and visual performances of the RTDs are exhibited, respectively, in **Table 2** and in **Figure 2a** and **b**. From **Table 2**, a very diverse pattern emerges in terms of mean and standard deviation with estimated volatilities ranging from 1.0 (NASDAQ) to 101.0 (Banda). Moreover, skewness appears mild everywhere, barring two cases where it is relatively high (S&P500 and NASDAQ), whereas kurtosis hovers for all RTDs around its critical value of three. All but two of the RTDs (AMO and SSN) are nonstationary and all exhibit zero valid regime switches, exclusion made for Banda and NASDAQ. Finally, they are all nonlinear except for AMO and SPI, as subsumed from the Harvey linearity test statistic whose critical value is close to 3.0. Optimal smoothing of the RTDs is achieved in the last two cases (Banda and NASDAQ) by means of subsample cluster analysis whereas for the other RTDs full-sample computation is more efficient (Section 2). The SRT2 smoother is found to exhibit the smallest PRMSEs in 75% of the cases proposed, whereas only three of them prefer the HHT method (S&P500, NASDAQ and SSN).

For each RTD, the actual signal, its optimal smoother (HHT or SRT2) and its linear trend are exhibited in **Figure 2a** and **b**. In **Figure 1**, horizontal (vertical) measurements represent time

Signal/ Statistic	Number of observations	Mean	Standard deviation	Skewness	Kurtosis	Harvey linearity test statistic	Corrected stationarity test statistic	Number of valid regime switches
AMO	1915	-0.030	0.241	0.046	3.499	2.994	-4.107	0
GISS	1627	1.508	32.994	0.608	2.698	1627.0	27.299	0
Yamalia	356	-0.714	1.790	-0.421	3.230	355.90	5.408	0
S&P500	789	478.650	550.361	1.121	2.977	789.000	83.865	0
SPI	1407	-0.233	1.039	-0.334	2.995	1.751	1.825	0
Banda	3000	584.863	59×10^3	0.279	2.896	3000.0	21.821	2
NASDAQ	536	1267.57	1263.04	1.048	3.299	536.0	38.445	3
SSN	315	54.365	40.457	0.663	2.440	315.000	-4.748	0

Table 2. Descriptive and select test statistics of the real-time dataset (RTD).

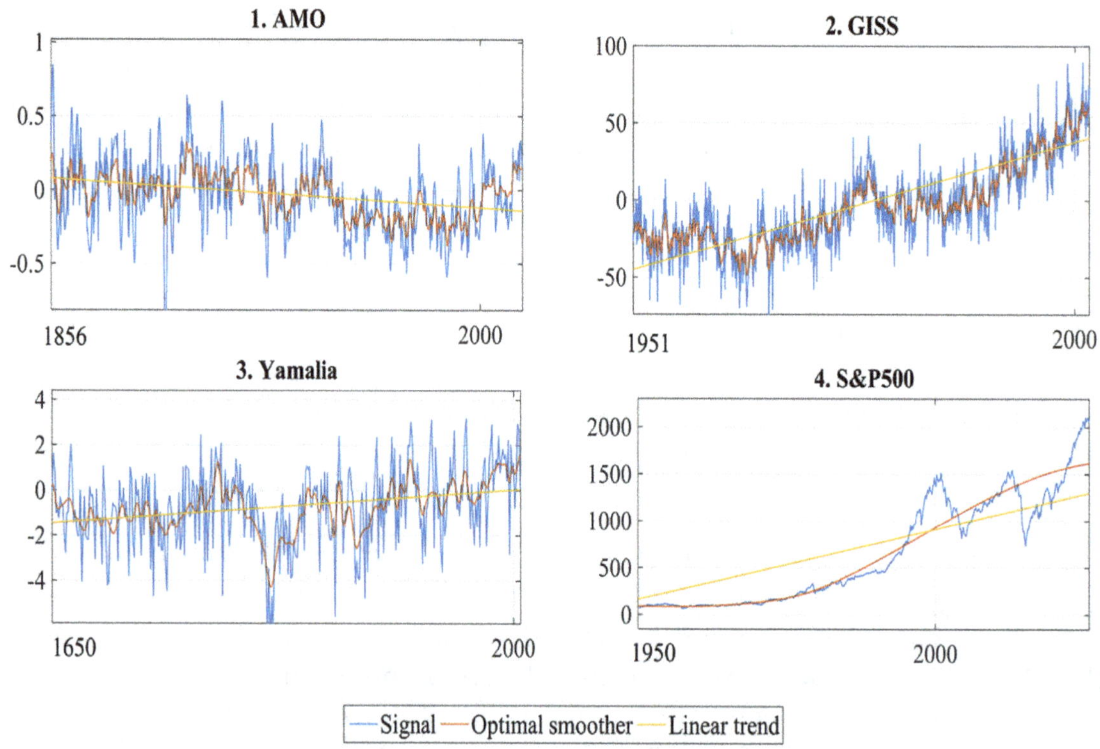

(a)

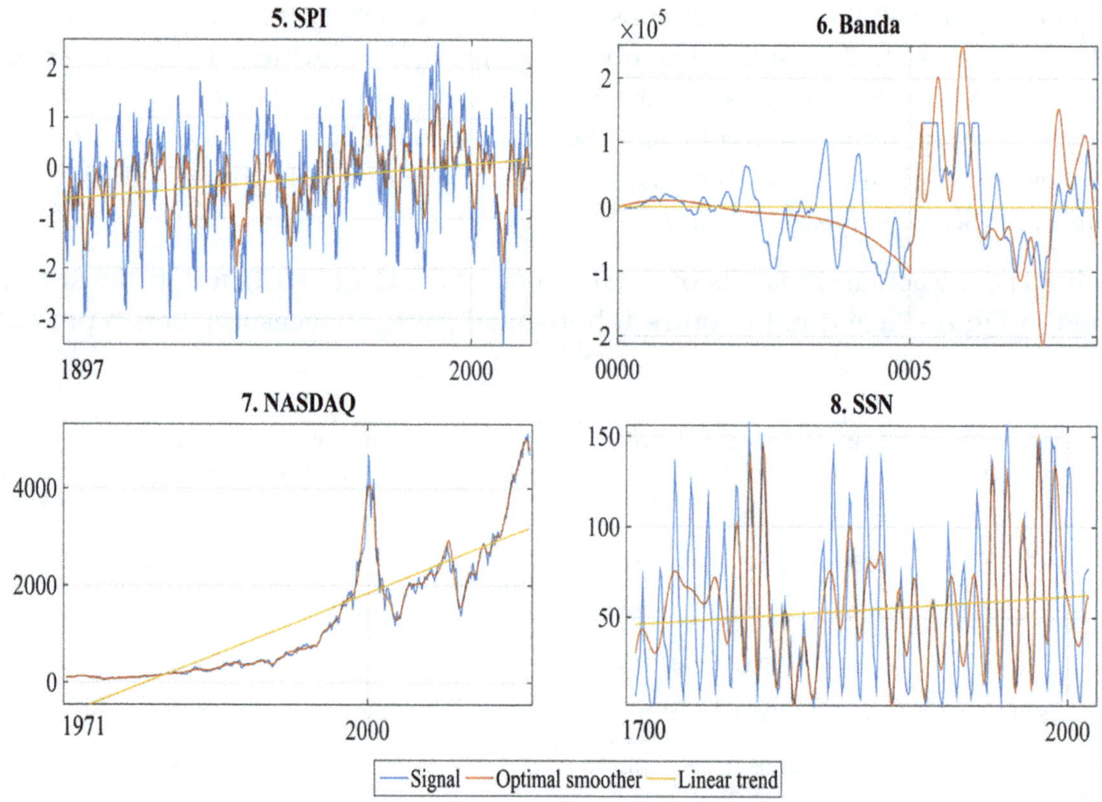

(b)

Figure 2. Real-time select random signals (a) vertical axes represent signal magnitudes, horizontal axes represent time in calendar years) and (b) vertical axes represent signal magnitudes, horizontal axes represent time in years (5, 7 and 8) or seconds (6).

lengths (magnitudes). Among the three HHT optimal smoothers found above, two are related to quirky signals, S&P500 and NASDAQ (**Figure 2a**, panel 4 and **Figure 2b**, panel 7). They exhibit however different frequencies, in spite of being estimated by the same subsample cluster technique. In fact the former (latter) is a low- (high-) resolution signal. Both signals exhibit in any case broken trends characterized by a long period of quiet followed by wild gyrations, which reflect the highly varying moods of both the stock market and of the Federal Reserve Board. The third signal associated with optimal smoothing attained through the HHT is the time series of sunspots (SSN). Full-sample estimation of the smoother was preferred, while broken trends clearly emerge from visual inspection (**Figure 2b**, panel 8). Significant regime switches are found to occur at the observations 1810 and 1902, in occasion of the Dalton and of the Modern Minimum, respectively [12].

Among the other five RTDs whose optimal smoother is of the SRT2 brand, only one (Banda) requires subsample cluster computation. The optimal smoother exhibits a dramatic regime switch in the passage from foreshocks to the main shocks and somewhat later (**Figure 2b**, panel 6). The two major break dates were found to be placed at observations 1833 and 1905 of 3000 observations. All of the RTD optimal smoothers, including Banda, exhibit high resolution and, at least visually, manifest considerable accuracy in tracking the original signal. This is particularly true for AMO, GISS, Yamalia and SPI, which are sizably noise-ridden (**Figure 2a**, panels 1–3 and 5).

6. Conclusions

This chapter has introduced and described in detail two new dual empirical methods for obtaining optimal smoothing of random signals pertaining to three broad taxonomic classes. Both methods utilize an application of the spectral representation theorem for signal decomposition that exploits the dynamic properties of optimal control. The two methods, named SRT1 and SRT2, produce a low- and a high-resolution filter, which may be utilized for optimal long- and short-run tracking as well as forecasting devices. The methods proven by Monte Carlo simulation found to be more efficient than the empirical Hilbert-Huang transform (HHT) for all of the taxonomic classes. The methods are also comparatively tested by using random artificial and a bunch of real-time signals, particularly eight select data sets including climatic, seismological and economic time series. HHT is proven to be more efficient in few cases of quirky, multiple regime-switch signals, like the Standard & Poor's 500 and the NASDAQ indexes.

Real-Time Data

1. AMO: Atlantic Multidecadal Oscillation, monthly observations 1856:01-2015:02. Available from: http://www.esrl.noaa.gov/psd/data/timeseries/AMO/ [Accessed: 2016-09-10].
2. GISS: Global Land-Ocean Temperature Index in 0.01 degrees Celsius, base period: 1951-1980, monthly observations 1880:01-2015:07. Sources: GHCN-v3 1880-07/2015 + SST: ERSST

v4 1880-07/2015. Available from: http://global-land-ocean-temperature-index.blogspot.com/p/blog-page_14.html [Accessed: 2015-08-20].

3. Yamalia Tree Ring Summer Temperature Reconstructions, yearly observations, 1650-2005. Available from: ftp://ftp.ncdc.noaa.gov/pub/data/paleo/treering/reconstructions/asia/russia/yamalia2013temp2000yr.txt [Accessed: 2016-09-10].

4. S&P500: Standard and Poor's Index, monthly observations adjusted close 1950:01-2015:09. Source: http://finance.yahoo.com/q/hp?s=%5EGSPC+Historical+Prices [Accessed: 2016-09-11].

5. SPI, U.S. Standardized Precipitation Index, monthly, 1897:01-2014:03. Available from: http://www.drought.gov/drought/content/products-current-drought-and-monitoring-drought-indicators/standardized-precipitation-index [Accessed: 2016-09-11].

6. Banda, Banda Aceh Earthquake, Indonesia, 12-26-2004, observations 12,000-14,999. Source: [41]. Available from: http://earthquake.usgs.gov/earthquakes/dyfi/events/us/ 2004-slav/us/ [Accessed: 2016-09-10].

7. NASDAQ Composite index, monthly close, 1971:02-2015:03. Available from: http://finance.yahoo.com/q/hp?s=%5EIXIC&a=01&b=5&c=1971&d=02&e=14&f=2015&g=m [Accessed: 2016-07-11].

8. SSN: Sun Spot Number Revisited Series by Clette et al., 2014, yearly 1700–2014. Available from: http://www.sidc.be/silso/datafiles [Accessed: 2016-09-11].

Author details

Guido Travaglini

Address all correspondence to: jay_of_may@yahoo.com

University of Rome 1, Rome, Italy

References

[1] Savitzky A. and Golay M.J.E. Smoothing and differentiation of data by simplified least squares procedures. Analytical Chemistry. 1964; **36**: 1627–1639. DOI: 10.1021/ac60214a047

[2] Daubechies I. Ten Lectures on Wavelets. SIAM, Society for Industrial and Applied Mathematics,Philadelphia,PA.1992. p.xix+357. DOI:10.1137/1.9781611970104;10.1137/1.9781611970104#_blank#Opens new window

[3] Hodrick R. and Prescott E.C. Postwar U.S. business cycles: an empirical investigation. Journal of Money, Credit and Banking. 1997; **29**: 1–16.

[4] Huang N.E., Shen Z., Long S.R., Wu M.C., Shih H.H., Zheng Q., Yen N.-C., Tung C.C. and Liu H.H. The empirical mode decomposition and the Hilbert spectrum for nonlinear

and nonstationary time series analysis. Proceedings of the Royal Society in London. 1998; **454A**: 903–993. DOI: 10.1098/rspa.1998.0193

[5] Huang N.E., Wu M.-L., Qu W., Long S.R., Shen S.S.P., and Zhang J.E. Applications of Hilbert–Huang transform to non-stationary financial time series analysis. Applied Stochastic Models in Business and Industry. 2003; **19**: 245–268. DOI: 10.1002/asmb.506

[6] Huang N.E. and Wu Z. A review on Hilbert-Huang transform: method and its applications to geophysical studies. Review of Geophysics. 2008; **46**: RG2006. DOI: 10.1029/2007RG000228

[7] Vivó-Truyols G. and Schoenmakers P.J. Automatic selection of optimal Savitzky-Golay smoothing. Analytical Chemistry. 2006; **78**: 4598–4608. DOI: 10.1021/ac0600196

[8] Xiang Z. and Peyman M. Automatic parameter selection for denoising algorithms using a no-reference measure of image content. IEEE Transactions on Image Processing. 2010; **19**: 3116–3132. DOI: 10.1109/TIP.2010.2052820

[9] Baxter M. and King R.G. Measuring business cycles: approximate band-pass filters for economic time series. Review of Economics and Statistics. 1999; **81**: 575–593. DOI: 10.3386/w5022

[10] Ravn M.O. and Uhlig H. On adjusting the Hodrick-Prescott filter for the frequency of observations. Review of Economics and Statistics. 2002; **84**: 371–376.

[11] McElroy T. Exact formulas for the Hodrick Prescott filter. The Econometrics Journal. 2008; **11**: 209–217. DOI: 10.1111/j.1368-423X.2008.00230.x

[12] Travaglini G. An econometric investigation of the sunspot number record since the year 1700 and its prediction into the 22nd century. Advances in Space Research. 2015; **56**: 992–1002. DOI: 10.1016/ j.asr. 2015. 05.012.

[13] Cramér H. On some classes of nonstationary stochastic processes. Proceedings of the Fourth Berkeley Symposium on Mathematical Statistics and Probability, Berkeley and Los Angeles, University of Los Angeles Press. 1961, pp. 57–77.

[14] Ng S. and Perron P. Lag length selection and the construction of unit root tests with good size and power. Econometrica. 2001; **69**: 1519–1554. DOI: 10.1111/1468-0262.00256

[15] Su J.-J., Cheung A.W.-K. and Roca E. Lag selection of the augmented Kapetanios-Shin-Snell nonlinear unit root test. Journal of Mathematics and Statistics. 2013; **9**: 102–111. DOI: 10.3844/jmssp.2013

[16] Rappoport P. and Reichlin L. On broken trends, random walks and non-stationary cycles. Technological and Social Factors in Long-Term Fluctuations. Di Marco M., et al. editors, Siena, IT. Springer-Verlag. 1989, pp. 305–331. DOI: 10.1007/978-3-642-48360-816

[17] Nason G.P. Stationary and nonstationary time series. Ch. 11, Statistics in Volcanology. Mader H. and Coles S.C. editors. The Geological Society, London. 2006, pp. 129–142.

[18] Harvey D.I., Leybourne S.J. and Xiao B. A powerful test for linearity when the order of integration is unknown. Studies in Nonlinear Dynamics & Econometrics. 2008; **12:** Article 2. DOI: 10.2202/1558-3708.1582

[19] Said E. and Dickey D.A. Testing for unit roots in autoregressive moving average models of unknown order. Biometrika. 1984; **71**: 599–607. DOI: 10.1093/biomet/71.3.599

[20] Kapetanios G., Shin Y. and Snell A. Testing for a unit root in the nonlinear STAR framework. Journal of Econometrics. 2003; **112**: 359–379. DOI: 10.1016/S0304-4076(02)00202-6

[21] Kapeitanos G. and Shin Y. Testing the null hypothesis of nonstationary long memory against the alternative hypothesis of a nonlinear ergodic model. Econometrics Review. 2011; **30**: 620–645. DOI: 10.1080/07474938.2011.553568 DOI:10.1080/07474938.2011.553568#_self

[22] Perron P. The great crash, the oil price shock and the unit root hypothesis. Econometrica. 1989; **57**: 1361–1401. DOI: 10.2307/1913712

[23] Bai J. and Perron P. Computation and analysis of multiple structural change models. Journal of Applied Econometrics. 2003; **18**: 1–22. DOI: 10.1002/jae.659

[24] Kim D. and Perron P. Unit root tests allowing for a break in the trend function at an unknown time under both the null and alternative hypotheses. Journal of Econometrics. 2009; **148**: 1–13. DOI: 10.1016/j.jeconom.2008.08.019

[25] Hamilton J.D. A new approach to the economic analysis of nonstationary time series and the business cycle. Econometrica. 1989; **57**: 357–384. DOI: 10.2307/1912559

[26] Hamilton J.D. Time Series Analysis. Princeton University Press, Princeton, NJ. 1994, pp. xiv + 799. DOI: 10.1017/S0266466600009440; 10.1017/S0266466600009440#_blank

[27] Hamilton J.D. Regime switching models. The New Palgrave Dictionary of Economics, 2nd ed., Durlauf S.N. and Blume L.E. editors. Palgrave Macmillan, London. 2008. DOI: 10.1057/9780230226203.1411

[28] Kanas A. Purchasing power parity and Markov regime switching. Journal of Money Credit and Banking. 2006; **38**: 1669–1687. DOI: 10.1353/mcb.2006.0083

[29] Mizrach B. Nonlinear time series analysis. The New Palgrave Dictionary of Economics, 2nd ed., Durlauf S.N. and Blume L.E. editors. Palgrave Macmillan, London, UK. 2008, pp. 169–177.

[30] Mizrach B. Nonlinear mean reversion in EMS exchange rates. Brussels Economic Review. 2010; **53**: 187–198.

[31] Lee H.-T. and Yoon G. Does purchasing power parity hold sometimes? Regime switching in real exchange rates. Applied Economics. 2013; **45**: 2279–2294. DOI: 10.1080/00036846.2012661399

[32] Wu Z., Huang N.E., Long S.R. and Peng C.-K. On the trend, detrending and variability of nonlinear and nonstationary time series. PNAS. 2007; **104**: 14889–14894. DOI: 10.1073/pnas.0701020104

[33] Wu Z. and Huang N.E. Ensemble empirical mode decomposition: a noise-assisted data analysis method. Advances in Adaptive Data Analysis. 2009; **1**: 1–41. DOI: 10.1142/S1793536909000047

[34] Dätig M. and Schlurmann T. Performance and limitations of the Hilbert-Huang transformation (HHT) with an application to irregular water waves. Ocean Engineering. 2004; **31**: 1783–1834.

[35] Jolliffe I.T. A note on the use of principal components in regression. Applied Statistics. 1982; **31**: 300–303.

[36] Jolliffe I.T. Principal Component Analysis, 2nd ed., Springer-Verlag, New York. 2002, pp. xxix + 427.

[37] Ahn S.C. and Horenstein A.R. Eigenvalue ratio test for the number of factors. Econometrica. 2013; **81**: 1203–1227. DOI: 10.3982/ECTA8968

[38] Miller R.E. Dynamic Optimization and Economic Applications. McGraw-Hill, New York, NY. 1979, pp. vii + 332.

[39] Athans M. and Falb P. Optimal Control: An Introduction to the Theory and Its Applications. Dover Publications, New York. 2007, pp. xvi + 879.

[40] Ross I.M. A Primer on Pontryagin's Principle in Optimal Control. Collegiate Publishers, San Francisco, CA. 2009, p. 102.

[41] IRIS, Incorporated Research Institutions for Seismology, 2015. Available from: https:/ds.iris.edu.ds.nodes.dmc [Accessed: 2016-09-11].

Permissions

All chapters in this book were first published in ASMTARP, by InTech Open; hereby published with permission under the Creative Commons Attribution License or equivalent. Every chapter published in this book has been scrutinized by our experts. Their significance has been extensively debated. The topics covered herein carry significant findings which will fuel the growth of the discipline. They may even be implemented as practical applications or may be referred to as a beginning point for another development.

The contributors of this book come from diverse backgrounds, making this book a truly international effort. This book will bring forth new frontiers with its revolutionizing research information and detailed analysis of the nascent developments around the world.

We would like to thank all the contributing authors for lending their expertise to make the book truly unique. They have played a crucial role in the development of this book. Without their invaluable contributions this book wouldn't have been possible. They have made vital efforts to compile up to date information on the varied aspects of this subject to make this book a valuable addition to the collection of many professionals and students.

This book was conceptualized with the vision of imparting up-to-date information and advanced data in this field. To ensure the same, a matchless editorial board was set up. Every individual on the board went through rigorous rounds of assessment to prove their worth. After which they invested a large part of their time researching and compiling the most relevant data for our readers.

The editorial board has been involved in producing this book since its inception. They have spent rigorous hours researching and exploring the diverse topics which have resulted in the successful publishing of this book. They have passed on their knowledge of decades through this book. To expedite this challenging task, the publisher supported the team at every step. A small team of assistant editors was also appointed to further simplify the editing procedure and attain best results for the readers.

Apart from the editorial board, the designing team has also invested a significant amount of their time in understanding the subject and creating the most relevant covers. They scrutinized every image to scout for the most suitable representation of the subject and create an appropriate cover for the book.

The publishing team has been an ardent support to the editorial, designing and production team. Their endless efforts to recruit the best for this project, has resulted in the accomplishment of this book. They are a veteran in the field of academics and their pool of knowledge is as vast as their experience in printing. Their expertise and guidance has proved useful at every step. Their uncompromising quality standards have made this book an exceptional effort. Their encouragement from time to time has been an inspiration for everyone.

The publisher and the editorial board hope that this book will prove to be a valuable piece of knowledge for researchers, students, practitioners and scholars across the globe.

List of Contributors

Agnar Höskuldsson
Centre for Advanced Data Analysis, Denmark

Roqueline Ametila e Gloria Martins de Freitas Aversi-Ferreira
Nursing and Pharmacy School, FAPAL, Tocantins, Brazil

Hisao Nishijo
System Emotional Science, Department of Physiology, School of Medicine and Pharmaceutical Sciences, University of Toyama, Toyama, Japan

Tales Alexandre Aversi-Ferreira
Institute of Biomedical Sciences, University of Alfenas, MG, Brazil

Aris Spanos
Department of Economics, Virginia Tech, Blacksburg, VA, USA

Indranil Ghosh
Department of Mathematics and Statistics, University of North Carolina, Wilmington, Wilmington, NC, USA

Gholamhossein G. Hamedani
Department of Mathematics, Statistics and Computer Science, Marquette University, Milwaukee, WI, USA

Sean Schmidt and Suzanna Long
Department of Engineering Management and Systems Engineering, Missouri University of Science and Technology, Rolla, MO, USA

Toshio Uchiyama
Hokkaido Information University, Ebetsu-shi, Hokkaido, Japan

Yves Dominicy
Université libre de Bruxelles, SBSEM, ECARES, Brussels, Belgium

Corinne Sinner
Université libre de Bruxelles, Département de Mathématique, Brussels, Belgium

Hong Son Hoang and Remy Baraille
SHOM/HOM/REC, Toulouse, France

Graham V. Weinberg
National Security, Intelligence, Surveillance, Reconnaissance Division, Defence Science, Technology Group, Edinburgh, South Australia, Australia

Rafdzah Zaki
Department of Social & Preventive Medicine, Faculty of Medicine, Julius Centre University of Malaya, Kuala Lumpur, Malaysia

Guido Travaglini
University of Rome 1, Rome, Italy

Index

www.ingramcontent.com/pod-product-compliance
Lightning Source LLC
LaVergne TN
LVHW060558210926
848485LV00005B/1334